Communications in Computer and Information Science 563

Commenced Publication in 2007
Founding and Former Series Editors:
Alfredo Cuzzocrea, Dominik Ślęzak, and Xiaokang Yang

More information about this series at http://www.springer.com/series/7899

José-Luis Sierra-Rodríguez · José Paulo Leal
Alberto Simões (Eds.)

Languages, Applications and Technologies

4th International Symposium, SLATE 2015
Madrid, Spain, June 18–19, 2015
Revised Selected Papers

 Springer

Editors
José-Luis Sierra-Rodríguez
Complutense University of Madrid
Madrid
Spain

Alberto Simões
Universidade do Minho
Braga
Portugal

José Paulo Leal
Universidade do Porto - DCC
Porto
Portugal

ISSN 1865-0929   ISSN 1865-0937   (electronic)
Communications in Computer and Information Science
ISBN 978-3-319-27652-6   ISBN 978-3-319-27653-3   (eBook)
DOI 10.1007/978-3-319-27653-3

Library of Congress Control Number: 2015957052

This Springer imprint is published by SpringerNature
The registered company is Springer International Publishing AG Switzerland

Preface

This volume contains the revised and extended proceedings of the fourth edition of SLATE, the 4th Symposium on Languages, Applications and Technologies, held at the Complutense University of Madrid, Spain, during June 18–19, 2015.

We continually use languages. First, to communicate between ourselves. Later, to communicate with computers. And more recently, with the advent of networks, we found a way to make computers communicate between themselves. All these different forms of communication use languages, different languages, but that still share many similarities. In SLATE we are interested in discussing these languages. Languages being such a broad subject, SLATE is organized in three main tracks:

- HHL Track: Processing Human–Human Languages. HHL is a forum dedicated to the discussion of research projects and ideas involving natural language processing and their industrial application. In 2015 this track was chaired by Alberto Simões.
- HCL Track: Processing Human–Computer Languages. HCL is a forum where researchers, developers, and educators exchange ideas and information on the latest academic or industrial work on language design, processing, assessment, and applications. The SLATE 2015 HCL chair was José-Luis Sierra-Rodríguez.
- CCL Track: Processing Computer–Computer Languages. The main goal of CCL is to provide a broad platform for discussion on the XML markup language: examples of usage and associated technologies. In 2015 this track was chaired by José Paulo Leal.

In this 4th edition we received 40 full-paper submissions and seven short-paper submissions. After a thorough peer-review process, in which each paper was reviewed by three anonymous reviewers, 17 papers were accepted as full papers (about 42 % full-paper acceptance rate). In addition, ten full papers were invited to be resubmitted as short papers, and nine short papers were finally accepted for publication and presentation at the symposium (about 52 % short-paper acceptance rate). This volume contains the extended and revised versions of all the papers presented at SLATE 2015.

The presentations were divided into the following eight sessions: Document Processing (CCL), Domain-Specific Languages (HCL), Tools for Natural Language Speech and Text Processing (HHL), Web Technologies and Practical Cases (CCL), Semantic and Text Classification (HHL), Human–Computer Language Processing (HCL), Semantic Web and Ontologies (CCL), and Grammars (HCL). In addition, a Lightning Talks and Demo Session was also included allowing for late-breaking presentations and tool demos. Finally, the SLATE 2015 program also included two keynotes: one on the application of grammar inference to software language engineering, by Marjan Mernik from the University of Maribor, Slovenia, and another on the role of ontologies in machine for machine communication, by Asunción Gómez from the Technical University of Madrid, Spain.

The organizers of SLATE 2015 want to thank the many people without whom this event would never have been possible. In particular, the UCM's Faculty of Philology for serving as the venue of the symposium; the UCM's Computer Science School for sponsoring the keynotes through its PhD Program Conference Cycle; the UCM's General Foundation for being in charge of the finances of the event; the Madrid City Council for organizing the symposium social activities; the Santander-UCM research grants for partially funding one of the keynotes; the members of the ILSA (Implementation of Language-Driven Software and Applications) Research Group for their collaboration in the local organization of the event; Springer for giving us the opportunity to publish this proceedings volume as part of the CCIS series; COMLAN and COMSIS journals for accepting submissions of additionally revised and extended journal-oriented versions of the best papers presented at the symposium; the EasyChair conference management system; the Program Committee members for spending their time reviewing the papers and writing the reports; the authors of the submitted papers for their contribution and interest in the symposium; and, finally, to all participants who came to Madrid and contributed such a fruitful meeting.

October 2015

José-Luis Sierra-Rodríguez
José Paulo Leal
Alberto Simões

Organization

Program Chairs

General Chair

José-Luis Sierra-Rodríguez

HCL Track Chair

José-Luis
  Sierra-Rodríguez

Universidad Complutense de Madrid

HHL Track Chair

Alberto Simões

Universidade do Minho

CCL Track Chair

José Paulo Leal

Universidade do Porto

HHL Program Committee

Alberto Simões   Universidade do Minho, Portugal
António Teixeira   Universidade de Aveiro, Portugal
Brett Drury   Universidade de São Paulo, Brazil
Fernando Baptista   Instituto Universitário de Lisboa, Portugal
Hugo Gonçalo Oliveira   Universidade de Coimbra, Portugal
Jörg Tiedemann   Uppsala Universitet, Sweden
Jorge Baptista   Universidade do Algarve, Portugal
José João Almeida   Universidade do Minho, Portugal
Lluís Padró   Universitat Politècnica de Catalunya, Spain
Octavian Popescu   IBM, T.J. Watson Research Center, NY, USA
Pablo Gamallo   Universidade de Santiago de Compostela, Spain
Thiago Pardo   Universidade de São Paulo, Brazil
Ulrich Heid   Universität Stuttgart, Germany
Xavier Gómez Guinovart   Universidade de Vigo, Spain

HCL Program Committee

Alda Lopes Gançarski   Institut National des Télécommunications, France
António Menezes Leitão   Universidade Técnica de Lisboa, Portugal
Bostjan Slivnik   Univerza v Ljubljani, Slovenia
Casiano Rodriguez-Leon   Universidad de La Laguna, Spain

Daniela da Cruz	Universidade do Minho, Portugal
Ivan Lukovic	University of Novi Sad, Serbia
Guido Wachsmuth	Delft University of Technology, The Netherlands
Jan Kollar	Technical University of Kosice, Slovakia
Jan Janousek	Czech Technical University, Czech Republic
Jaroslav Poruban	Technical University of Kosice, Slovakia
Jean-Cristophe Filliâtre	Laboratoire de Recherche en Informatique, France
João Paiva Cardoso	Universidade do Porto, Portugal
José-Luis Sierra-Rodríguez	Universidad Complutense de Madrid, Spain
Josep Silva	Universidad Politécnica de Valencia, Spain
Maria João Varanda Pereira	Instituto Politécnico de Bragança, Portugal
Mario Beron	Universidad Nacional de San Luis, Argentina
Marjan Mernik	Univerza v Mariboru, Slovenia
Nuno Oliveira	Universidade do Minho, Portugal
Nuno Ramos	Universidade do Minho, Portugal
Paulo Matos	Instituto Politécnico de Bragança, Portugal
Pedro Rangel Henriques	Universidade do Minho, Portugal
Ricardo Rocha	Universidade do Porto, Portugal
Salvador Abreu	Universidade de Évora, Portugal
Simão Melo de Sousa	Universidade da Beira Interior, Portugal

CCL Program Committee

Alda Lopes Gançarski	Institut National des Télécommunications, France
Alexander Paar	TWT GmbH Science and Innovation, Germany
Cristina Ribeiro	Universidade do Porto, Portugal
Daniel Diaz	University of Paris 1 - Pantheon Sorbonne, France
Eugenijus Kurilovas	Centre of Information Technology in Education, Lithuania
Gabriel David	Universidade do Porto, Portugal
Giovani Librelotto	Universidade Federal de Santa Maria, Brazil
João Correia Lopes	Universidade do Porto, Portugal
José Carlos Ramalho	Universidade do Minho, Portugal
José Paulo Leal	Universidade do Porto, Portugal
Pedro Rangel Henriques	Universidade do Minho, Portugal
Peter Sloep	Open Universiteit, The Netherlands
Ricardo Queirós	Instituto Politécnico do Porto, Portugal
Salvador Abreu	Universidade de Évora, Portugal

Organizing Committee

Pedro Rangel Henriques	Universidade do Minho, Portugal
José Paulo Leal	Universidade do Porto, Portugal
Alberto Simões	Universidade do Minho, Portugal

Maria João Varanda	Instituto Politécnico de Bragança, Portugal
José-Luis Sierra-Rodríguez	Universidad Complutense de Madrid, Spain
Antonio Sarasa-Cabezuelo	Universidad Complutense de Madrid, Spain
Antonio Pareja-Lora	Universidad Complutense de Madrid, Spain
Ana Fernandez-Pampillon	Universidad Complutense de Madrid, Spain
Daniel Rodriguez-Cerezo	Universidad Complutense de Madrid, Spain

Additional Reviewers

Antonio Navarro	Universidad Complutense de Madrid, Spain
Antonio Sarasa	Universidad Complutense de Madrid, Spain
Félix Buendía	Universidad Politécnica de Valencia, Spain
Helena Moniz	INESC-ID, Portugal
Maha Khemaja	ISSATSo - University of Sousse, Tunisia
Marcos García	Universidade de Santiago de Compostela, Spain
Miguel Anxo Solla Portela	Universidade de Vigo, Spain
Vicente Blanco	Universidade de La Laguna, Spain

Contents

Computer-Computer Languages

Human-Human Languages

Speech Features for Discriminating Stress Using Branch and Bound Wrapper Search

Mariana Julião[1]([✉]), Jorge Silva[1], Ana Aguiar[1], Helena Moniz[2,3], and Fernando Batista[2,4]

[1] Instituto de Telecomunicações,
Rua Dr. Roberto Frias, s/n, 4200-465 Porto, Portugal
{mjuliao,ana.aguiar}@fe.up.pt, up201007483@alunos.dcc.fc.up.pt
http://www.it.pt
[2] INESC-ID, Lisboa, Portugal
{helenam,fmmb}@l2f.inesc-id.pt
[3] FLUL/CLUL, Universidade de Lisboa, Lisboa, Portugal
[4] ISCTE - Instituto Universitário de Lisboa, Lisboa, Portugal

Abstract. Stress detection from speech is a less explored field than Automatic Emotion Recognition and it is still not clear which features are better stress discriminants. The project VOCE aims at doing speech classification as stressed or not-stressed in real-time, using acoustic-prosodic features only. We therefore look for the best discriminating feature subsets from a set of 6125 features extracted with openSMILE toolkit plus 160 Teager Energy Operator (TEO) features. We use a Mutual Information (MI) filter and a branch and bound wrapper heuristic with an SVM classifier to perform feature selection. Since many feature sets are selected, we analyse them in terms of chosen features and classifier performance concerning also true positive and false positive rates. The results show that the best feature types for our application case are Audio Spectral, MFCC, PCM and TEO. We reached results as high as 70.4 % for generalisation accuracy.

Keywords: Stress · Emotion recognition · Ecological data · Feature sets · Feature selection

1 Introduction

The motivations for detecting stress from speech range from it being a non-intrusive way to detect stress, to ranking emergency calls [7], or improve speech recognition systems, since it is known that environmentally induced stress leads to fails on speech recognition systems [13]. Public Speaking is said to be "the most common adult phobia" [18], showing the relevance of a tool to improve public speaking. In VOCE[1], we target developing such a tool, by developing algorithms to identify emotional stress from live speech. In particular, VOCE

---

[1] http://paginas.fe.up.pt/~voce.

© Springer International Publishing Switzerland 2015
J.-L. Sierra-Rodríguez et al. (Eds.): SLATE 2015, CCIS 563, pp. 3–14, 2015.
DOI: 10.1007/978-3-319-27653-3_1

corpus comes mainly from public speaking events that occur within academic context, like presentations of coursework or research seminars. The envisioned coaching application requires detecting emotional stress in live speech in near real time, to give the user timely feedback, which requires adapting the computational costs to the limited memory and computational resources to use. Decreasing the number of features used for classification reduces the amount of data to collect, the amount of features to be extracted and the complexity of the classifier, impacting a reduction in the memory and computational resources used. Additionally, feature selection can increase the classifier's accuracy [12]. Thus, in this paper, we focus on identifying these reduced feature sets based on their performance as stress discriminators.

In this work, we start from the fusion of two feature sets: the group of features extracted using the openSMILE toolkit [25], and the group of TEO-based features, to be detailed on Sect. 4.2. We filter these feature sets with Mutual Information (MI) and then use a branch-and-bound wrapper to explore the space of possible feature sets. Finally, we analyse the best feature sets chosen on various branches for the most frequently chosen feature categories.

2   Related Work

The importance of suprasegmental acoustic phenomena that can be taken as global emotion features is highlighted in [28], like "hyper-clear speech, pauses inside words, syllable lengthening, off-talk, disfluency cues, inspiration, expiration, mouth noise, laughter, crying, unintelligible voice". These features have been mainly annotated by hand, and automatic extraction is not straightforward, though possible in some cases.

Stress recognition from speech is a specific case of emotion recognition. The Fundamental Frequency, F0, is the most consensual feature for stress discrimination [8,14,22,31], but several metrics for energy and formant changes have been proposed, often represented by Mel-Frequency Cepstral Coefficients (MFCCs) [7,21,31]. Frequency and amplitude perturbations – Jitter and Shimmer –, and other measures of voice quality, like Noise to Harmonics Ratio and Subharmonics to Harmonics Ratio [26,28] have also been used. Teager Energy Operator-based features have also shown to perform well in speech under stress [31], and we shall look at them in detail in this work.

TEO-based features have been shown to increase recognition robustness with car noise [10,15]. In [17], TEO-based features reached the best performance for stressed speech discrimination outdoor, but not indoor. They also have been used to do voiced-unvoiced classification [19]. In the latter work, the advantages of TEO are enunciated: because only three samples are needed for the energy computation at each time instant, it is nearly instantaneous. Therefore, this time resolution allows to capture energy fluctuation, and also a robust AM-FM estimation in noisy environment. [6] uses Teager Energy Operator in the development of a system for hypernasal speech detection. In this work, we shall look into the discrimination power of TEO-based speech features for stress discrimination in public speaking.

3 Speech Corpus and Data Annotation

The VOCE corpus [2] currently consists of 38 raw recordings from voluntaries aged 19 to 49. Data is recorded in an ecological environment, concretely during academic presentations[2]. Speech was automatically segmented into utterances, according to a process described in [5].

Annotation into stressed or neutral classes was performed per speaker, based on the mean heart rate [4]. Utterances on the third quartile of mean heart rates for that speaker are annotated as stressed, while the remaining ones are annotated as neutral.

Using an ecologically collected corpus imposes an unavoidable trade-off between the quality of the recording and the spontaneity of the speaker. Higher quality of the recording not only allows for more reliable feature extraction, in general, but also impacts the performance of the segmentation algorithms we use to split the speech into sentence-like units – utterances –, and to do text transcription, necessary for the extraction of TEO features. For these reasons, we chose only 21 raw recordings for this work.

For these speakers, 1457 valid utterances were obtained[3]. The set of utterances is divided into 15 speakers (507 utterances) for training and 6 speakers

Table 1. Dataset demographic data. PSE: Public Speaking Experience, 1 – 5: 1 - little experience, 5 - large experience.

Train set				Test set			
Age	Gender	PSE	#Utts	Age	Gender	PSE	#Utts
26	male	2	56	24	male	3	97
22	male	2	39	19	male	2	61
24	male	3	36	19	male	3	86
21	male	3	38	19	female	3	64
22	male	3	32	23	female	4	71
22	male	3	25	19	female	3	63
25	male	2	54				
19	male	3	12				
21	male	3	22				
21	female	3	51				
24	female	5	27				
22	female	2	37				
21	female	3	32				
21	female	3	18				
19	female	3	28				

---

[2] Please refer to [3] for details on the collection methodology.
[3] Remaining utterances after discarding 94 utterances with length of less than 1 s or more than 25 s.

(442 utterances) for testing. Since the number of stressed utterances corresponds to approximately 1/4 of the total, we randomly down-sampled the train data in order to balance the two classes, which led to the mentioned 507 utterances. During feature selection, the classifier was trained on 354 utterances and tested on 153 utterances. These utterances belonged to the train set. Table 1 characterises the dataset concerning age, gender, public speaking experience, and the number of utterances considered[4].

We performed outlier detection on each feature using the Hampel identifier [20] with t = 10. The outliers were then replaced by the mean value of the feature excluding outliers, and feature values were scaled to the interval [0,1].

4   Methodology

Figures 1(a) and (b) illustrate the workflow for speech segmentation and feature selection, respectively. In this work, we start from the fusion of two feature sets: the group of features extracted using the openSMILE toolkit [25], and the group of TEO-based features, to be detailed on Sect. 4.2. We filter these feature sets with Mutual Information and then use a branch-and-bound wrapper to explore the space of possible feature sets. We then analyse the best feature sets chosen on various branches for the most frequently chosen feature categories.

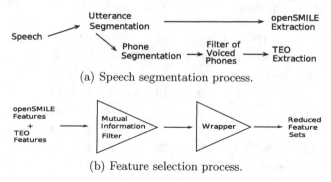

(a) Speech segmentation process.

(b) Feature selection process.

Fig. 1. Workflow for the speech segmentation and the feature selection process.

4.1   Acoustic-Prosodic Features

OpenSMILE extracts a set of 128 low-level features at the frame level from the speech signal, known as low-level descriptors (LLD) [11]. Statistical functionals are then applied over the LLD in order to compute values for longer segments, providing a total of 6125 features at the segment level [25]. These features and their extraction processes are described in [9,24].

---

[4] Please note that the stated number of utterances on the train set corresponds to the one actually used after discarding a part of the neutral utterances, and not to the number of utterances in the natural set.

The openSMILE toolkit is capable of extracting a very wide range of acoustic-prosodic features and has been applied with success in a number of paralinguistic classification tasks [23]. It has been used in the scope of this study to extract a feature vector containing 6125 speech features, by applying segment-level statistics (means, moments, distances) over a set of energy, spectral and voicing related frame-level features.

4.2 Teager Energy Operator Features

The following TEO-Based features were extracted: Normalized TEO autocorrelation envelope and Critical Band Based TEO Autocorrelation Envelope as in [31]. The literature where Normalized TEO Autocorrelation Envelope and Critical Band Based TEO Autocorrelation are presented targets the feature extraction for small voiced parts usually called "tokens" [31]. To work equivalently, we did a phone recognition with the delimitation of each phone [1] and used only voiced sounds. These correspond to phones represented by the portuguese SAMPA symbols 'i', 'e', 'E', 'a', '6', 'O', 'o', 'u', '@', 'i~', 'e~', '6~', 'o~', 'u~', 'aw', 'aj', '6~j~', 'v', 'z', 'Z', 'b', 'd', 'g', 'm', 'n', 'J', 'r', 'R', 'l', 'L' [29, Chap. IV.B].

These features are extracted per frame. The length of each frame is about 10ms, depending on the feature to extract. Each phone usually contains many frames and each utterance has normally many phones. Therefore, since we want to have values per utterance, we consider each feature extracted for all phones and apply statistics to it. These statistics are: mean, standard deviation, skewness, kurtosis, first quartile, median, third quartile, and inter-quartile range. This process is also illustrated in Fig. 1(a). The first two columns in Table 2 summarise the feature types considered in this work[5].

5 Searching for the Best Feature Sets

As already stated, we apply one filter to reduce the dimensionality from initially 6285 functional (OS) plus TEO features before applying the wrapper with a Support Vector Machine (SVM) classifier with radial basis function kernel and C=100[6], using *python* library *scikit-learn*.

5.1 Filter: Mutual Information

There are several metrics and algorithms to compute the relevance of features on a dataset, and the choice of this metric may hugely impact the final subset of features. However, since there is a lack of a priori knowledge about filter metric adequacy to specific datasets [30], we based our choice on the work of Sun and

---

[5] The generic designation "type" is the result of aggregating Low Level Descriptor features with their derived functionals (e.g., quartiles, percentiles, means, maxima, minima). This procedure is, in our perspective, a way to better group and interpret the performance of the features.

[6] This value was found empirically to produce the best classification results.

Table 2. Feature Types: Id, Name, Number of features of each type selected for MI, Number of features of each type chosen for the Best Sets: T.A.1, T.A.2, G.A., Se., Sp., and Comb.

Id	Type	MI	T.A.1	T.A.2	G.A.	Se.	Sp.	Comb.
1	F0final[a]	10	0	0	0	0	0	0
2	TEO[b]	17	3	2	1	3	2	3
3	audSpec_Rfilt[b]	187	8	8	6	7	6	6
4	audspec[a]	6	0	0	0	0	0	0
5	audspecRasta[a]	4	0	0	0	0	0	0
6	jitterDDP	6	0	0	1	0	0	0
7	jitterLocal[a]	7	0	0	0	0	0	0
8	logHNR	8	1	0	0	0	1	1
9	mfcc_sma[b]	119	7	8	4	2	3	4
10	pcm_Mag_fband	17	0	0	1	2	1	0
11	pcm_Mag_harmonicity	14	1	0	0	0	0	0
12	pcm_Mag_psySharpness[b]	6	3	2	1	1	1	2
13	pcm_Mag_spectralEntropy	6	0	1	0	0	0	1
14	pcm_Mag_spectralFlux	8	0	1	0	0	0	0
15	pcm_Mag_spectralKurtosis[a]	7	0	0	0	0	0	0
16	pcm_Mag_spectralRollOff	22	1	0	0	0	0	0
17	pcm_Mag_spectralSkewness[a]	1	0	0	0	0	0	0
18	pcm_Mag_spectralSlope	6	0	1	1	0	0	0
19	pcm_Mag_spectralVariance[a]	10	0	0	0	0	0	0
20	pcm_RMSenergy	6	1	1	1	1	1	0
21	pcm_zcr	8	0	0	0	1	0	0
22	shimmerLocal[a]	8	0	0	0	0	0	0
23	voicingFinalUnclipped[a]	4	0	0	0	0	0	0

[a] Type not selected by the best sets
[b] Type always selected by the best sets

Li et al. [27], which showed good results in terms of classification for Mutual Information (MI), a metric that measures the mutual dependence between two random variables.

Since MI is based on the probability distribution of discrete variables and our features have continuous values, we had to define a binning. We (1) defined five binning possibilities: 50, 100, 250, 500 or 1000 bins; (2) computed MI for each feature and each binarisation possibility; (3) kept features for which the MI value belonged to the higher quartile for all binarisation options. Their distribution per feature type corresponds to the third column in Table 2.

5.2   Wrapper

Feature selection has been widely studied and, as result, a large number of algorithms have been proposed. These algorithms can be categorized into three groups: filter, wrapper and embedded [16]. Wrapper algorithms find the final solution using a learning algorithm as part of the evaluation criteria. The main idea of these methods is to use the learning algorithm as a black-box to guide the search for the optimal solution. The learning is applied to every candidate solution and the goodness of the subset is given according to the performance of the learning algorithm. Due to the learning algorithm being directly used on the process of selecting features, these methods tend to find better solutions. Nonetheless, the final solution only applies for the selected learning algorithm, since using a different one will most likely result on a different final solution. These methods have higher computational cost as they require training and classifying data for each candidate solution.

We designed a branch and bound wrapper to search the space of feature sets obtained from the MI filter for the combination of features that deliver the best classifier performance. This wrapper starts by searching all combinations of sets up to 10 features, keeping all that are within 1.5 % accuracy of the best solution found so far. Larger feature sets are obtained by expanding the previously kept solutions with blocks of features not yet in the sets. Every time a feature subset is tested with a classification algorithm, a score is produced, which is the accuracy, in this case. Subsets are kept and expanded if the expansion improves the previous accuracy. This search runs until the work list of feature sets with new combinations empties. This wrapper provides a better exploration of the feature set space than traditional forward and backward wrappers. Since the search space for our wrapper is much bigger than for most wrapper methods, we used parallel programming techniques to improve the throughput of the algorithm, using *python'sMultiprocessing* package.

6   Results

The Mutual Information filter selected 487 features, distributed into types as described in the third column of Table 2. After choosing the best 280 feature sets with training accuracies below 85 % from 20 processors, we looked at their distribution by feature types, which is on Fig. 2.

Among these 280 feature sets we looked for the ones having the best scores in each of the considered metrics[7]: Train Accuracy, Generalisation Accuracy, Sensitivity (Se), Specificity (Sp)[8], and a Combined Metric defined as

$$\text{CombinedMetric} = \frac{(\text{Se} + \text{Sp})}{2} - |\text{Sp} - \text{Se}|. \qquad (1)$$

---

[7] Generalisation Accuracy, Sensitivity and Specificity are computed on the test set.

[8] Being TP - number of True Positives, TN - number of True Negatives, FP - number of False Positives, FN - number of False Negatives, Sensitivity=$\frac{\text{TP}}{\text{TP+FN}}$ and Specificity=$\frac{\text{TN}}{\text{TN+FP}}$.

The need for this metric follows from the fact that it is our goal not only to have a good generalisation accuracy, but also to have high sensitivity and high specificity at the same time. This is relevant since, as we have an imbalanced test set, with much more neutral utterances than stressed utterances, it can happen that high generalisation results are due to high values of true positives, while true negatives are neglected – which is the kind of scenario we want to avoid. On Table 3, each line corresponds to the best feature subset for which the metric specified in the first column was found to be maximum. The two last lines correspond to baseline results, meaning the classification for the whole set of features and for the set of MI filtered features.

Columns T.A.1, T.A.2, G.A., Se., Sp., and Comb, in Table 2, correspond to the best feature sets, according to each of these metrics, as exposed in Table 3. Each of the Columns in Table 2 signs the number of features of each type (each line corresponds to a feature type).

Table 3. Metrics for the Best Subsets as percentage

Set	Train Acc.	Gen. Acc.	Sens.	Spec.	Comb.	# features
Train Acc	84.97	61.76	59.81	62.39	58.53	25
Train Acc	84.97	62.22	52.34	65.37	45.82	24
Gen. Acc	81.70	70.36	33.64	82.09	9.42	16
Sensitivity	81.70	59.28	71.96	55.22	46.85	17
Specificity	81.05	70.14	31.78	82.39	6.47	15
Combined	81.70	64.03	61.68	64.78	60.14	17
Complete	—	63.12	50.47	67.16	42.13	6285
MI	—	60.86	45.79	65.67	35.85	487

Table 3 bears the following information:

- The sets of best train accuracy do not correspond to the ones with best generalisation accuracy. Actually, these have the second worse generalisation results among these sets.
- The set of best generalisation accuracy, as well as the set of best specificity, although having very good generalisation accuracies have very low sensitivities. This is the kind of imbalance we want to avoid.
- The same train accuracy can have sets of very different quality. We see that for train accuracy 81.70 % we have the best generalisation accuracy, the best sensitivity and the best combined metric. Looking at the other columns in the table we see that only the line for Combined Metric has acceptable results in sensitivity and specificity.
- *These best reduced sets often achieve better results than both the complete set and the filtered set, having much smaller sets, which is very good for the envisioned real-time public speaking coaching application.*

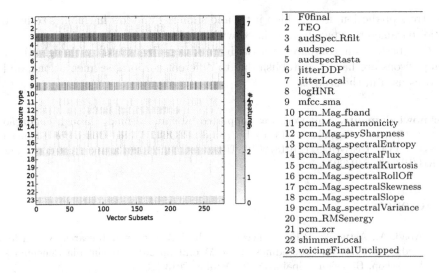

1	F0final
2	TEO
3	audSpec_Rfilt
4	audspec
5	audspecRasta
6	jitterDDP
7	jitterLocal
8	logHNR
9	mfcc_sma
10	pcm_Mag_fband
11	pcm_Mag_harmonicity
12	pcm_Mag_psySharpness
13	pcm_Mag_spectralEntropy
14	pcm_Mag_spectralFlux
15	pcm_Mag_spectralKurtosis
16	pcm_Mag_spectralRollOff
17	pcm_Mag_spectralSkewness
18	pcm_Mag_spectralSlope
19	pcm_Mag_spectralVariance
20	pcm_RMSenergy
21	pcm_zcr
22	shimmerLocal
23	voicingFinalUnclipped

Fig. 2. Heatmap for feature type frequencies on each subset.

7 Discussion

The set of features selected by the Mutual Information filter are, *grosso modo*, the ones reported in the literature for other languages (e.g., [14,32]). Those encompass pitch information, mostly final movements of pitch, audio spectral differences, voice quality features (jitter, shimmer, and harmonics-to-noise-ratio) and TEO features, the latter usually described as very robust across gender and languages. As for PCMs and MFCCs, these features are very transversal in speech processing tasks and highly informative for a wide range of tasks, not surprising, thus, for stress detection as well. The features selected by Mutual Information filter give us a more complete characterization of stress predictors. From these set the ones that are systematically chosen in the best features sets using the wrapper are mostly TEO, MFCCs and audio spectral differences. TEO and MFCCs features are also reported by [32], for English and Mandarin, as the most informative ones, even more than pitch itself.

8 Conclusions

We have used a corpus of ecologically collected speech to search for the best speech features that discriminate stress. Starting from 6125 features extracted with openSMILE toolkit and 160 Teager Energy features, we used a Mutual Information filter to obtain a reduced subset for stress detection. Next, we searched for the best feature set using a branch and bound wrapper with SVM classifiers.

Our results provide further evidence that the features resulting from the Mutual Information filtering process are robust for stress detection tasks, independently of the language, and highlight the importance of voice quality features for stress prediction, mostly high jitter and shimmer and low harmonics to noise ratio, parameters typically associated with creaky voice.

Our best result compares well with work done by [10,32], although direct comparisons are hard to establish due to different corpora, segmentations, and metrics used in the studies.

Acknowledgments. This work was supported by national funds through Fundação para a Ciência e Tecnologia (FCT) by project VOCE (Voice Coach for Reduced Stress) PTDC/EEA-ELC/121018/2010, UID/CEC/50021/2013, and Post-doc grant SFRH/PBD/95849/2013.

References

1. Abad, A., Astudillo, R.F., Trancoso, I.: The L2F spoken web search system for mediaeval 2013. In: Proceedings of the MediaEval 2013 Multimedia Benchmark Workshop, Barcelona, Spain, 18–19 October 2013 (2013)
2. Aguiar, A., Kaiseler, M., Meinedo, H., Almeida, P., Cunha, M., Silva, J.: VOCE corpus: ecologically collected speech annotated with physiological and psychological stress assessments. In: Calzolari, N., Choukri, K., Declerck, T., Loftsson, H., Maegaard, B., Mariani, J., Moreno, A., Odijk, J., Piperidis, S. (eds.) Proceedings of the Ninth International Conference on Language Resources and Evaluation (LREC 2014). European Language Resources Association (ELRA), Reykjavik (2014)
3. Aguiar, A.C., Kaiseler, M., Meinedo, H., Abrudan, T.E., Almeida, P.R.: Speech stress assessment using physiological and psychological measures. In: Mattern, F., Santini, S., Canny, J.F., Langheinrich, M., Rekimoto, J. (eds.) UbiComp (Adjunct Publication), pp. 921–930. ACM (2013)
4. Allen, M.T., Boquet, A.J., Shelley, K.S.: Cluster analyses of cardiovascular responsivity to three laboratory stressors. Psychosom. Med. **53**(3), 272–288 (1991)
5. Batista, F., Moniz, H., Trancoso, I., Mamede, N.J.: Bilingual experiments on automatic recovery of capitalization and punctuation of automatic speech transcripts. IEEE Trans. Audio Speech Lang. Process. **20**(2), 474–485 (2012)
6. Cairns, D.A., Hansen, J.H.L., Kaiser, J.F.: Recent advances in hypernasal speech detection using the nonlinear teager energy operator. In: ICSLP 1996, p. 1 (1996)
7. Demenko, G.: Voice stress extraction. In: Proceedings of the Speech Prosody 2008 Conference (2008)
8. Demenko, G., Jastrzebska, M.: Analysis of voice stress in call centers conversations. In: Proceedings of Speech Prosody, 6th International Conference, Shanghai, China (2012)
9. Eyben, F., Wllmer, M., Schuller, B.: openSMILE: the munich versatile and fast open-source audio feature extractor. In: Bimbo, A.D., Chang, S.F., Smeulders, A.W.M. (eds.) ACM Multimedia, pp. 1459–1462. ACM (2010)
10. Fernandez, R., Picard, R.W.: Modeling drivers' speech under stress. Speech Commun. **40**(1–2), 145–159 (2003)
11. Ferreira, J., Meinedo, H.: VOCE project stress feature survey technical report 2. Technical report, L2F, Inesc-ID, Lisboa, Portugal, November 2013
12. Guyon, I., Elisseeff, A.: An introduction to variable and feature selection. J. Mach. Learn. Res. **3**, 1157–1182 (2003)

13. Hansen, J.H., Bou-Ghazale, S.E., Sarikaya, R., Pellom, B.: Getting started with the susas: A speech under simulated and actual stress database. Technical Report: RSPL-98-10 (1998)

14. Hansen, J.H., Patil, S.A.: Speech under stress: Analysis, modeling and recognition (2007)

15. Jabloun, F., Cetin, A.E., Erzin, E.: Teager energy based feature parameters for speech recognition in car noise. IEEE Sig. Process. Lett. **6**, 259–261 (1999)

16. Kumar, V., Minz, S.: Feature selection: a literature review. Smart CR **4**(3), 211–229 (2014)

17. Lu, H., Frauendorfer, D., Rabbi, M., Mast, M.S., Chittaranjan, G.T., Campbell, A.T., Gatica-Perez, D., Choudhury, T.: Stresssense: detecting stress in unconstrained acoustic environments using smartphones. In: Proceedings of the 2012 ACM Conference on Ubiquitous Computing, UbiComp 2012, pp. 351–360. ACM, New York (2012). http://doi.acm.org/10.1145/2370216.2370270

18. Miller, T.C., Stone, D.N.: Public speaking apprehension (psa), motivation, and affect among accounting majors: a proofofconcept intervention. Issues Account. Educ. **24**(3), 265–298 (2009)

19. Sundaram, N., Smolenski, B., Yantorno, R.: Instantaneous nonlinear teager energy operator for robust voicedunvoiced speech classification (2003)

20. Pearson, R.K. (ed.): Exploring Data in Engineering, the Sciences, and Medicine. Oxford University Press, USA (2011)

21. Sarikaya, R., Gowdy, J.N.: Subband based classification of speech under stress. In: ICASSP, pp. 569–572 (1998)

22. Scherer, K.R., Grandjean, D., Johnstone, T., Klasmeyer, G., Bnziger, T.: Acoustic correlates of task load and stress. In: Hansen, J.H.L., Pellom, B.L. (eds.) INTERSPEECH. ISCA (2002)

23. Schuller, B., Steidl, S., Batliner, A., Burkhardt, F., Devillers, L., MüLler, C., Narayanan, S.: Paralinguistics in speech and language-state-of-the-art and the challenge. Comput. Speech Lang. **27**(1), 4–39 (2013)

24. Schuller, B., Batliner, A., Seppi, D., Steidl, S., Vogt, T., Wagner, J., Devillers, L., Vidrascu, L., Amir, N., Kessous, L., Aharonson, V.: The relevance of feature type for the automatic classification of emotional user states: low level descriptors and functionals. In: INTERSPEECH, pp. 2253–2256. ISCA (2007)

25. Schuller, B., Steidl, S., Batliner, A., Nöth, E., Vinciarelli, A., Burkhardt, F., van Son, R., Weninger, F., Eyben, F., Bocklet, T., Mohammadi, G., Weiss, B.: The interspeech 2012 speaker trait challenge. In: INTERSPEECH. ISCA (2012)

26. Sun, X.: A pitch determination algorithm based on subharmonic-to-harmonic ratio. In: the 6th International Conference of Spoken Language Processing, pp. 676–679 (2000)

27. Sun, Z., Li, Z.: Data intensive parallel feature selection method study. In: 2014 International Joint Conference on Neural Networks (IJCNN), pp. 2256–2262, July 2014

28. Vogt, T., André, E., Wagner, J.: Automatic recognition of emotions from speech: a review of the literature and recommendations for practical realisation. In: Peter, C., Beale, R. (eds.) Affect and Emotion in Human-Computer Interaction. LNCS, vol. 4868, pp. 75–91. Springer, Heidelberg (2008)

29. Wells, J.: Handbook of Standards and Resources for Spoken Language Systems. Mouton de Gruyter, Berlin (1997)

30. Wolpert, D.H.: The lack of a priori distinctions between learning algorithms. Neural Comput. **8**(7), 1341–1390 (1996)

31. Zhou, G., Hansen, J., Kaiser, J.: Nonlinear feature based classification of speech under stress. IEEE Trans. Speech Audio Process. **9**, 201–216 (2001)
32. Zuo, X., Fung, P.N.: A cross gender and cross lingual study on acoustic features for stress recognition in speech. In: Proceedings 17th International Congress of Phonetic Sciences (ICPhS XVII), Hong Kong, pp. 2336–2339 (2011)

Oriya Morphological Analyzer Using Lttoolbox

Itisree Jena, Himani Chaudhry$^{(\boxtimes)}$, and Dipti Misra Sharma

International Institute of Information Technology, Hyderabad, India
{itisree,himani}@research.iiit.ac.in, dipti@iiit.ac.in

Abstract. A Morphological analyzer is an essential tool for many NLP applications. Developing a fully-fledged morphological analyzer (MA) tool for an agglutinative language like Oriya is a challenging task. This paper deals with development of a MA for Oriya, a resource poor language. The MA is being developed using the paradigm approach. It consists of various paradigms under which nouns, pronouns, adjectives, verbs and indeclinables are classified. The paradigms have been created for inflected forms using an XML based morphological dictionary from the Lttoolbox package. At present, a total of 10,840 words have been entered into the dictionary. In the course of the paper we talk about the design and implementation of the MA. We also talk about the issues and limitations experienced in developing it using Lttoolbox.

Keywords: Oriya morphological analyzer · Apertium · Lttoolbox · Compound verb · Paradigm approach · Morphological analyzer

1 Introduction

Efforts are on for past many years, to develop various of NLP tools such as Morphological Analyzers (MA), Part-of-Speech (POS) taggers, spell checkers and so on, for Indian languages (ILs), to assist tasks such as Machine Translation. These efforts to develop NLP tools for ILs have especially focused on computational morphology, since ILs are morphologically quite rich. Developing an Oriya MA becomes important, to help build these tools for the language. This work presents the design and development of a MA for Oriya.

The official language of the state of Odisha (Orissa), Oriya, now officially pronounced 'Odia' belongs to the eastern branch of the Indo Aryan sub family of the Indo-European language. It has the status of the sixth classical language in India. Around 31 million people are using this language. "Oriya is a syntactically head-final and morphologically agglutinative language" [15]. Thus, quite some information is contained in morphological structures in Oriya.

The nouns in Oriya are generally characterized by inflectional categories like number, gender, case and also take articles and number classifiers. The definite articles '-ti' and '-taa' occur only with singular nouns. Its plural markers include '-maane', '-gudaa', '-gudika', '-gudaaka'. The plural marker '-maane' is only added to animate nouns e.g. 'pua+maane (son+s). It can not be added to either human proper nouns or in-animate nouns. Thus, we can not say 'kaatha+maane' (wood+s).

© Springer International Publishing Switzerland 2015
J.-L. Sierra-Rodríguez et al. (Eds.): SLATE 2015, CCIS 563, pp. 15–25, 2015.
DOI: 10.1007/978-3-319-27653-3_2

Oriya has natural gender that does not reflect in the agreement with grammatical categories like verbs. For example, baagha (tiger) and baaghuNi (tigress). We use roman transliteration scheme to represent examples in this paper.

In Oriya, "adjectives which precede the nouns in attributive position do not show any agreement with the nouns except in a few cases where the adjective agrees with the noun in gender" [12]. For example, kaLaa baLada (black bull), kaaLi gaaii (black cow).

Oriya finite verbs are marked for person, number, tense, aspect and mood. They agree with subject nouns and this is reflected by an agreement marker that manifests attached to the end of the main verb. For example:

```
aame khaa-il-u
we   eat-pst-agr.
'We ate'
```

1.1 Related Work

Various methods have been adopted for morphological analysis in natural language processing. Brute Force method, Root Driven approach, Affix Stripping method are some of the methods evolved typically for the analysis of ILs. MAs being developed using the paradigm approach include Hindi MA by Bharati et al. (1995) [3], and the Marathi MA by Bapat et al. (2010) [2], of which, [2] combine a paradigm based inflectional system with finite state machines for modeling the morphotactics. Marathi derivational MA by Vaidya et al. (2009) [16], Tamil MA by Parameshwari (2011) [13] and Benagli by Faridee et al. (2009) [7], all adopt paradigm approach using Lttoolbox to develop their MA, which is similar to our work, discussed later in this paper.

Further, Oriya MA have been developed by Shabadi (2003) [15], Sahoo (2003) [14] using deterministic Finite State Automata (FSA), where the FSA recognize if the input string of morphemes is an appropriate Oriya word or not. They do this by plugging each forms into the FSA, using two level morphology. The work propose a model which can provide lexical, morphological and syntactic information for each lexical unit in the analyzed word form. The second approach followed for Oriya is our work using Lttoolbox from the Apertium toolkit which we have reported in Jena et al. (2011) [8].

2   Current Work

2.1 Approach

We have adopted the paradigm based approach to create a MA for Oriya. Paradigms are employed to represent the inflectional regularities of lexical units in a language [6]. A paradigm is a set of related word forms which follow the same set of spelling rules and take the same kind of affixes. "Paradigm approach is well suited for agglutinative language nature" [1]. Oriya being an agglutinative language, the paradigm approach seems to work well for it.

2.2 Resources Used

Lexical Resources. The foremost requisite for a MA is a root word dictionary. But, Oriya being a resource poor language, an online root word dictionary wasn't available for it. We, thus, manually created the dictionary using the following resources:

- 'Taruna Sabdakosha' [9] an Oriya dictionary.
- 'A synchronic grammar of Oriya' [12].
- A corpus of 2,720,400 words from Central Institute of Indian Languages, Mysore (CIIL) - Our major resource for the database of the dictionary and also for the training and testing data for our MA.

The root word dictionary was created using the lexical resources mentioned above. Initially, we used a frequency based list from the CIIL Oriya corpus and added root words to it from the 'Taruna Sabdakosha', to enhance it. Currently the dictionary contains 10,840 root words, details of which are:

1. Nouns - 5,031
2. Pronouns - 18
3. Adjectives - 930
4. Verbs - 4,537
5. Adverbs - 179

6. Postpositions - 40
7. Conjunctions - 36
8. Clitics - 15
9. Particles - 14
10. Interjections - 40

Tool. We used Lttoolbox [11] package from the Apertium [6] toolkit to develop the Oriya MA. The Lttoolbox is a well known NLP tool used to build tools like morphological analyzer and morphological generator. It is a free software and released under the terms of the GNU General Public License. It uses an XML based format to represent linguistic data. Paradigms are created inside it using some of the elements in its morphological dictionary. Further, a morphological dictionary can be used for both, a morphological analyzer and a morphological generator, depending on the direction in which it is read by the system.

2.3 Data Development for Oriya Morph Analyzer

Oriya Morphological Dictionary in Lttoolbox. The Oriya morphological dictionary consists of declension or conjugation patterns of words in XML format used in Lttoolbox. The dictionary has four sections, of these the two main sections are paradigm definition section and dictionary section. Alphabet and symbol definition sections being the other two sections.

Declension or conjugation used, are based on parameters such as, gender, number, person, case, vibhakti (case marker) for nouns and pronouns. Gender, number, person, suffix string taken as TAM (tense, aspect and modality) for verbs. [3]

Classification of Paradigms. Paradigms have been created for the open class categories like nouns, verbs and adjectives and later on, closed class categories like postpositions and conjunctions etc. The words that have identical grammatical information make one paradigm class. However, all words with similar endings/suffixes may not follow the same paradigm. For instance, two verbs 'khaa' (eat) and 'gaa' (sing) fall in the same paradigm as they take similar inflections. But the verb 'jaa' (go) falls in a different paradigm though it has the same ending. This is because the verb 'jaa' (go) changes its root form when it takes past tense inflection e.g. 'jaa' (go) becomes 'gali' (go+past) but in case of verb 'gaa' (sing) becomes 'gaaili'(sing+past). There are some parts of speech like adverbs, conjunctions, postpositions, clitics etc., that remain uninflected, so we have listed them directly in our dictionary. Table 1 shows the paradigm classification for different categories.

Table 1. Number of paradigm classes.

S.No	Category	Paradigm classes
1	Noun	14
2	Pronoun	9
3	Adjective	13
4	Verb	13

3   Evaluation and Result

We conducted three experiments to evaluate our MA. We discuss this shortly. Since a MA produces more than one answer, we found it more appropriate to carry out a more detailed evaluation of the MA than just evaluating the precision and recall values, since "Precision-Recall gives general overall impression about the performance of a system" [10]. A more detailed evaluation is necessary to know what kind of words are over analyzed, which are under analyzed, and so on. This is discussed in detail, in Subsects. 3.3 and 3.4.

3.1   Evaluation I

Here we focus on the overall coverage of our MA (Table 2). A corpus of 11,368 words (non-unique) was taken (Sect. 2.2) in order to evaluate the overall coverage of the morph in a random test data environment.

It must be noted that the coverage here is based on a small dictionary size of 10,840 root words. The class of recognized words includes the cases where the tool gave an analysis (irrespective whether the analysis was correct, partially correct or wrong). While the class of unrecognized words comprises those cases where the morph analyzer didn't give an output or analysis.

Table 2. Results: the overall coverage.

Total no. of words	11,368
Recognized words	8,303
Unrecognized words	3,065
Coverage	73.03 %

3.2 Results and Error Analysis

In Table 2 we see that 3,065 words remained unrecognized by our MA, which forms 26.97 %. These words can be easily accounted for (Table 3 shows the break up of the unanalyzed words). Out of this 26.97 %, out of vocabulary (OOV) words (which include foreign words, proper nouns and numerals) form 29.81 % and noise (meaningless characters/words occurring in the corpus) takes up 6.62 %. The remaining words fall into causative verbs (2.34 %) and 'others' (61.20 %). Since causative verbs are currently not being handled, these remain unanalyzed. 'Others' in Table 3, are Oriya words that remain unanalyzed because they have yet to be entered in the morphological dictionary. These form a major part of the unrecognized words.

Therefore, the two major categories that affect the coverage of the MA, are OOV & noise (36.43 %) and 'Others' (61.20 %) of 3,065 unrecognized words. With a small dictionary size of 10,840 words, the MA's coverage is 73.03 % and increasing the dictionary size can further improve the coverage.

Table 3. Error analysis.

S.No	Unanalyzed words	Occurrences	%
1	Numerals	164	5.35
2	Foreign words	338	11.02
3	Proper nouns	412	13.44
4	Causative verbs	72	2.34
5	Others	1876	61.20
6	Noise	203	6.62
	Total	3065	100 %

3.3 Evaluation II

When an MA produces output, it may have 6 possible cases:

1. Type1: correct output, e.g. ABCD/ABCD.
2. Type2: added some wrong output to correct output, e.g. ABCD/ABCDE.
3. Type3: missed some correct output, e.g. ABCD/ABC.
4. Type4: missed some correct output and add some wrong output, e.g. ABCD/ ABCE.

5. Type5: all incorrect output, e.g. ABCD/EFG.
6. Type6: no output, ABCD/No Output.

These six cases help us to decide which aspect of morphology needs further attention for improvement. To evaluate an MA, some data manually tagged with morph features (gold-standard data) is needed. It contains all possible analysis of the words. In the above examples 'ABCD' is gold standard data and others are machine's output. To create the gold standard data to evaluate our MA we randomly took 1066 words from the CIIL Oriya corpus. The data was tagged using Sanchay (an open source platform for working on languages, with components like a text editor with customizable support for languages and encodings, annotation interfaces, etc.) annotation interface, in Shakti Standard Format (SSF) (This format is a highly readable representation for storing language analysis [4]). The Apertium produced morphological analysis was compared with the gold standard data.

We compared the machine produced morphological analysis using our gold standard data as the reference data. After we ran our MA on the randomly taken corpus we compared it with the gold standard data. Table 4 shows the results for type wise evaluation of the accuracy against a gold-standard corpus.

Table 4. Results: type wise evaluation of the accuracy against a gold-standard corpus.

S.No	Types	Gold/Output	Count	% Count
1	Type:1	ABCD/ABCD	754	70.73
2	Type:2	ABCD/ABCDE	41	3.84
3	Type:3	ABCD/ABC	25	2.34
4	Type:4	ABCD/ABCE	100	9.38
5	Type:5	ABCD/EFG	0	0
6	Type:6	ABCD/No Output	146	13.69
		Total tokens	1066	100 %

In Table 4, **Type:1** gives fully correct output (comprises 70.73 % of total count), whereas **Type:2**, **Type:3** and **Type:4** give partially correct output (comprises 15.56 % of the cases). Further, the coverage of the tool is 86.30 %. As mentioned earlier, **Type:4** consists of some correct output and some wrong output (partially correct output), we notice that **Type:4**–9.38 % has the highest contribution in cases with partially correct output, as compared to the other types with partially correct output (**Type:2**–3.84 % and **Type:3**–2.34 %). **Type:6** includes cases where MA fails to give the output.

3.4   Evaluation III

In the third evaluation we focused on the accuracy of only two features–'root' and 'category' instead of all of the features. This is so because for some applications only these two features are taken into consideration. Other feature structure values may not be important for them. Thus, through evaluation II the accuracy of the MA for such applications is also reported. Additionally, for evaluation III we took the same data sets that were used in evaluation II. Table 5 shows the results for type wise evaluation of the accuracy for two features.

Table 5. Results: Type wise evaluation of the accuracy for 'root' and 'category'.

S.No	Types	Gold/Output	Count	% Count
1	Type:1	ABCD/ABCD	853	80.01
2	Type:2	ABCD/ABCDE	21	1.96
3	Type:3	ABCD/ABC	27	2.53
4	Type:4	ABCD/ABCE	19	1.78
5	Type:5	ABCD/EFG	0	0
6	Type:6	ABCD/No Output	146	13.69
		Total tokens	1066	100 %

We see that the percentage count of **Type:1** increased to 80.01 % in evaluation II, whereas the percentage count of **Type:4** decreased to 1.78 % (dropped by 7.60 %). Thus considering only root and category features shows an overall higher accuracy of the MA. The coverage remains the same for both the evaluations.

4   Challenges and Limitations

4.1   Foreign Words

As seen in Sect. 3.2, foreign words remain unrecognized, and thus unanalyzed in our MA since they are not part of the data base. Presence of foreign words in ILs is a widely occurring phenomenon, given a high degree of code switching in ILs. They cause the coverage of the MA to go down. They are not a part of the Oriya morph dictionary since they are foreign language words and can not be included in the 'Oriya' dictionary.

A possible solution to handle these would be creating a separate dictionary for them. To add a separate tag 'foreign word' in the MA, a dictionary of foreign words would have to be included and manually created. However, since foreign words have widely occurring in the language, creating an exhaustive list would be required for the MA to tag them as 'foreign word'. This would be expensive in terms of time and resources. Further, though a work around for the problem, this is not a very good option either, as this would call for capturing too many irregularities by way of the inflections they take (or do not take). Capturing these irregularities falls out of the purview of our MA, as this would entail entering all these types of inflections in the dictionary. Since their taking of inflections is a productive process, this may make the task more complex, and may also fail generalization.

4.2 Analyzing Oriya Compound Verbs

Though simple verbs could be handled by creating paradigms for them, in Lttool-box with relative ease, handling Oriya compound verbs (CV) proved quite a challenge for us. Before we go on to discuss the issues we came across in this, we would like to discuss briefly about CV in general, and about Oriya CV in particular:

A Compound verb consists of two verbs (v1, v2), yet acts as a single verb. One of its components is a 'secondary' verb which carries inflections like gender, number, person, tense, aspects and modality and the other, the 'main' verb which carries most of the semantics of the compound, and determines its arguments. The 'secondary' verbs "cannot be said to be predicating fully, though they are clearly not entirely devoid of semantic predicative power" [5].

Forming compounds is a highly productive process in IL. In languages like Hindi and Oriya, secondary verbs are generally, a small set that form compounds with the 'main' verbs.

Structure and Behaviour of Oriya Compound Verbs. We have identified 13 'secondary' verbs in Oriya, as seen below in Table 6.

Table 6. List of 'secondary' verbs in Oriya.

S.No	Secondary verbs	Meaning
1	jaa	go
2	de	give
3	ne	take
4	pakaa	throw
5	bas	sit
6	paar	can
7	aas	come
8	pad	fall
9	uth	awake
10	chaal	walk
11	saar	finish
12	aaN	bring
13	he	happen

In Oriya CV the stem vowel '-i' attaches to the 'main' verb, which in turn is followed by a 'secondary verb' from a limited number of verb roots that occur as 'secondary' verbs. For example:

```
se so-i-pad-il-aa
he sleep-stemvowel-fall-pst-agr.
'He fell asleep'
```

The stem vowel '-i' is different from an aspectual marker, though both have/ take the same form. The difference between them is that the aspectual marker is followed only by an auxiliary verb while the non-aspectual marker which is a 'stem vowel' is followed by a secondary verb [12]. An example for this is:

```
a. se lekh-i-ch-i
   he write-prs-aux-agr.
   'He has written'
b. aame khaa-i-de-l-u
   we  eat-stemvowel-give-pst-agr.
   'we have finished eating'
```

Also, while in simple verbs inflectional suffixes attach to the main verb root, in CV the inflectional suffixes attach to the secondary verb, since Oriya is an agglutinative language, and these two verbs occur together. Thus the two verbs together arrive at a derived root. And so, we get two roots, a 'main' root and a 'derived' root.

Thus, since Oriya CV are different from simple verbs, in structure and behaviour, we can't analyze them like simple verbs even though they occur as a single entity most of the time. Also, since Oriya CV are composed of two verbs (v1, v2) agglutinated together, that arrive at a derived root, the output of our MA should give this inflectional information for each derived root, in order to capture the information about their structure and the derivation happening in it.

For example, in 'khaaidelaa' (finished eating), the root khaa 'eat' is where we get the information of the action 'eat'. When the secondary verb attaches to the main verb, another root 'khaaide' is derived. Our morph's output for the derived verb 'khaaidelaa' should thus be:

```
^khaaidelaa/root:khaa<droot:khaaide>
<dsuffix:de><cat:v><gen:any>
<num:sg><per:a><tam:ilaa>$
```

However, this may not be a feasible solution for us, since information pertaining to such output will have to be incorporated in the dictionary for each CV, making this a cumbersome task. Besides, since there wouldn't be any (scope of) generalization here, this would beat the purpose of using the paradigm approach.

Another solution for this would be the Apertium way–using nested paradigms to handle derivational forms, since Oriya CV are composed of combinations of verbs from the set of Oriya (main) verbs. "The use of nested paradigms is to facilitate the processes of derivation followed by inflection." [16]

Here, the paradigms of secondary verbs would be 'called' upon, from within the main verbs' paradigms to arrive at their compounds. For instance, the verb 'khaaideichi' is derived from the main verb 'khaa' (eat) and secondary verb 'de' (give) to form a compound. So the paradigm for 'de' is called from within the paradigm of the verb 'khaa'. Likewise, other secondary verbs would be 'called' from within the paradigm 'khaa' to form compounds.

However, not all verbs of a paradigm class take the same secondary verbs to form compounds. There are verbs that fall under the same paradigms (since they share same types of inflections) that form compounds with different sets of secondary verbs. For example, the verbs 'khaa' and 'gaa' are classified under the same paradigm class, but they take different secondary verbs. Thus, if we call all the secondary verbs that go with 'khaa', within the paradigm 'khaa', then while processing, the analyzer gives a similar output for 'gaa' also, though they don't take same secondary verbs. We say 'gaaiuthilaa' 'started singing (suddenly)' but we don't say 'khaaiuthilaa' 'started eating suddenly'. It thus leads to some ungrammatical structures also.

It needs a mention here, that though the nested paradigm approach may work for a MA, from the perspective of generation it may lead to generation of ungrammatical structures. Since these two modules are obtained from a single morphological dictionary (depending on the direction they are read from–left to right for analyzer and right to left for generator as given by [6], a different resolution is needed to resolve this.

Therefore, based on discussion above we conclude that using nested paradigms doesn't seem to be the best option for the analysis of Oriya CV.

```
<sdef n="droot:khaaide" c="khaaide"/>
```

The third, and a very simple approach to resolve this issue of handling Oriya CV in our MA, would be entering the derived roots of the CV in the morphological dictionary, in the dictionary section. The morphological dictionary contains the root or 'lemma', the part of the lemma which is common for all inflected forms, that is 'lemma cut' and the paradigm name. We simply add the derived root and the lemma cut of the derived root in the place where this information is entered in the dictionary. This would save us the task and the effort of preparing separate paradigms for the compound verbs.

For example, the dictionary entry for the CV 'khaaidelaa' (finished eating) with the derived root 'khaaide' would be:

```
<e lm="khaaide"><i>khaaid</i>
<par n="d/e__v"/></e>
```

< par > in the entry indicates which paradigm from among the ones defined in the < pardefs >, the derived root belongs to. Here, the derived root 'kaaide' falls under the paradigm for the root 'de', since the CV 'khaaidelaa' takes the inflections of the secondary verb 'de'. Thus reference to the 'de' paradigm through the element <par> saves us the effort of listing all the inflected forms of the derived root/lemma in the morphological dictionary entry.

The output our MA would give for the above example is:

```
^khaaide/khaaide<cat:v>
<gen:any><num:sg>
<per:m_h0><tam:imper>$
```

5 Conclusion and Future Work

In this paper we presented a paradigm based MA for Oriya using Lttoolbox from the Apertium toolkit. It is based on the concept of morphological paradigms. Currently it handles only inflectional morphology, and nouns, pronouns, adjectives, verbs, compound verbs and indeclinables have been included in its morphological dictionary. Since the MA is currently in its preliminary stage, addition of remaining categories and increasing the dictionary size for existing categories will improve its performance and increase its coverage. Using the Oriya MA for other NLP tools such as part of speech tagger, chunker, spell checker, machine translation system for Oriya can also be created in future. These would be a useful resource for the language.

References

1. Antony, P.J., Soman, K.P.: Computational morphology and natural language parsing for Indian languages: a literature survey. Int. J. Comput. Sci. Eng. Technol. 136–146 (2012)
2. Bapat, M., Gune, H., Bhattacharyya, P.: A paradigm-based finite state morphological analyzer for Marathi. In: 23rd International Conference on Computational Linguistics, pp. 26–34 (2010)
3. Bharati, A., Chaitanya, V., Sangal, R., Ramakrishnamacharyulu, K.: Natural Language Processing: A Paninian Perspective. Prentice-Hall, India (1995)
4. Bharati, A., Sangal, R., Sharma, D.: SSF: Shakti standard format guide. Technical report, IIIT Hyderabad (2007)
5. Butt, M.: The light verb jungle. In: Workshop on Multi-verb Constructions (2003)
6. Forcada, M., Bonev, B., Rojas, S., Ortiz, J., Sánchez, G., Martínez, F., Armentano-Oller, C., Montava, M., Tyers, F.: Documentation of the open-source shallow-transfer machine translation platform apertium (2008)
7. Faridee, A.Z.M., Tyers, F.M., Others.: Development of a morphological analyser for Bengali. Universidad de Alicante, Departamento de Lenguajes y Sistemas Informáticos (2009)
8. Jena, I., Chaudhury, S., Chaudhry, H., Sharma, D.M.: Developing Oriya morphological analyzer using Lt-toolbox. In: Singh, C., Singh Lehal, G., Sengupta, J., Sharma, D.V., Goyal, V. (eds.) ICISIL 2011. CCIS, vol. 139, pp. 124–129. Springer, Heidelberg (2011)
9. Kar, K.C.: Taruna Sabdakosha, vol. 1. Grantha Mandir, Cuttack (2000)
10. Kulkarni, A., Shukla, D.: Sanskrit morphological analyzer: some issues. In: Festschrift, B.K. (ed.) Volume by LSI (2009)
11. Lttoolbox. http://wiki.apertium.org/wiki/Lttoolbox
12. Mahapatra, B.P.: A Synchronic Grammar of Oriya. Udaya Narayana Singh, Central Institute of Indian Languages, Mysore (2007)
13. Parameshwari, K.: An implementation of APERTIUM morphological analyzer and generator for Tamil. In: Parsing in Indian Languages, p. 41 (2011)
14. Sahoo, K.: Oriya nominal forms: a finite state processing. In: TENCON 2003, Conference on Convergent Technologies for Asia-Pacific Region, vol. 2, pp. 730–734. IEEE (2003)
15. Shabadi, K.: Finite state morphological processing of Oriya verbal forms. In: Proceedings of EACL-2003 Workshop on Computational Linguistics for the Languages of South Asia: Expanding Synergies with Europe, pp. 49–56 (2003)
16. Vaidya, A., Sharma, D.: Using paradigms for certain morphological phenomena in Marathi. In: 7th International Conference on NLP (ICON-2009), pp. 132–139 (2009)

Exploiting Twitter for the Semantic Enrichment of Telecommunication Alarms

Hugo Gonçalo Oliveira[1]([⊠]), João Marques[1], and Luís Cortesão[2]

[1] CISUC, Department of Informatics Engineering,
University of Coimbra, Coimbra, Portugal
hroliv@dei.uc.pt, joliv@student.dei.uc.pt
[2] Portugal Telecom Inovação e Sistemas, Aveiro, Portugal
luis-m-cortesao@telecom.pt

Abstract. Everyday, several different alarms are triggered in a telecommunications network. Inspired by works that mine useful information from Twitter, we aim at exploiting this resource for semantically-enriching those alarms. We assume that, during the alarms, Twitter users would mention potential causes, and also that network customers would tweet to complain about the quality of their service. For this purpose, we explored a set of alarms and tweets from the same period of time and came to the conclusion that tweets on potential causes of the alarms are hard to find. The most significant findings are that, during an alarm, there are more tweets related to rain events, or those swearing and thus a sign of complaint.

Keywords: Information extraction · Event detection · Social network mining · Twitter · Telecommunication alarms

1 Introduction

The increasing popularity of social networks such as Twitter or Facebook has made social media a relevant part of people's lives. These networks are highly accessible and have hundreds of millions of users all over the world, who read, post, and share real-time messages, in a fast pace. They have thus become relevant sources of information, also suitable for exploitation by computational tools that acquire precious knowledge, such as people's opinion on certain subjects [11], current trends, or general events [3].

Inspired by works that mine useful information from Twitter (see Sect. 2), our project aims at exploiting this social network for semantically-enriching alarms triggered by technical problems in a telecommunications network. Ideally, Twitter would provide relevant information and contribute to a better understanding of the alarm's cause (e.g. natural disaster, accident, concentration of people), thus leading to additional measures by the network managers, to minimize negative consequences. It could be further used for analysing the impact of the alarms on the network customers (e.g. whether they lead to complaining tweets). To this

© Springer International Publishing Switzerland 2015
J.-L. Sierra-Rodríguez et al. (Eds.): SLATE 2015, CCIS 563, pp. 26–37, 2015.
DOI: 10.1007/978-3-319-27653-3_3

end, we have used all the alarms triggered during an entire month, for the wired and mobile networks of Portugal Telecom, one of the major telecommunication operators in Portugal. For the same period of time, we collected all the tweets we could obtain from Twitter's public streaming API, published in Portugal and written in Portuguese.

This paper reports on several experiments with the previous datasets, explored in order to gather more insights on the data and assess the suitability of Twitter for the task at our hands. These preliminary experiments involved the combination of the alarm data with the Twitter data, to identify tweets posted at the same time and from the same place of an alarm, as well as a shallow analysis of the tweets text, for further exploitation. Though we do not see this work as finished, the performed experiments have shown that we have a very challenging goal. In fact, looking for useful tweets has revealed to be the same as looking for a needle in a haystack. This might be due to the small population of Portugal, or to the lower popularity of Twitter in our country, as compared to other countries where experiments of this kind were quite successful. This adds to the small size of the sample of tweets that we can get for free and to the long term of some alarms. Despite all these issues, there is much to report, and we strongly believe that some of the presented results might be relevant for other researchers using Portuguese tweets for different purposes.

The remaining of the paper starts with an overview on information extraction from Twitter, with a focus on event detection. Following, we describe our dataset of alarms and tweets. After this, we present the experiments performed to investigate whether the alarms could be semantically enriched by Twitter, including a manual classification of tweets published during alarms according to their utility, searching for potentially relevant keywords, and classifying the tweets automatically, according to mentioned events. We end by speculating on possible reasons for the lack of useful tweets, and discuss future directions for this research, which should move on to explore other available data sources.

2   Background and Related Work

Twitter is a microblogging social network, with ≈288M monthly active users and 500M messages (tweets) sent every day[1]. This overwhelming number of tweets, available every second with fresh information, made Twitter an attractive media for research on text mining and information extraction (IE). But Twitter holds specific features that make it particular and increase complexity, resulting in poor performance by traditional tools. Tweets use informal language, with many abbreviations; they ignore some grammatical rules and conventions (e.g. they rarely use capital letters); are limited to 140 characters; and they use hashtags (#) to provide additional context. So, specific natural language processing (NLP) tools had to be developed for IE from Twitter (e.g. named entity recognizers [8]), used in tasks such as opinion mining [11] or event detection [3,12].

---

[1] Numbers according to https://about.twitter.com/company.

Close to our work, *TwiCal* [12] is a system that extracts event calendars from Twitter, based on probabilistic latent variable models. Events are characterised by their name, date, description, and type (e.g. sports, politics, meeting). *TwiCal* assumes that the involved entities and the event date play an important role (e.g. events in the same date tend to be of the same type). Based on similar assumptions on the date, event summaries have been produced [4]. To make tweets processing more efficient, there is also work on the summarization of tweets [20]. Other approaches to event detection cluster tweets according to their timestamp, location, hashtags and used text [3,19]. Clusters not related to an event are discarded.

The topic of event detection from Twitter has been applied to a varied range of more specific tasks, such as sub-event identification in football matches [1], crime prediction [18], user vacation plans and revealing medical conditions mining [10], disease rates and alcohol sales volumes [5], or to natural crisis management. On the latter, Twitter has been used for tracking forest fires [9], reporting [14], detecting [7] and assessing the damages [2] of earthquakes, or for the early detection of tsunamis [21]. Those systems exploit the real-time nature of Twitter and the fast spreading of information it provides. They also consider space and time information for detecting natural disasters. Yet, although the publication time is obtained in a straightforward manner, some researchers report that it does not always match the real-word spread of the disaster [9]. Identifying the event location might be even more problematic because not all tweets have explicit coordinates attached; their coordinates are not always very accurate; and tweets are sometimes posted from a different location than the disaster. This is why there is work on geo-tagging tweets with unavailable coordinates, with applications to crisis management [6].

Despite a close relation to the described works, we are not aware of Twitter mining for improving the description of alarms in a telecommunications network, only for events in general news [17]. To some extent, this is a specific application of using Twitter for gathering additional information on known events [4]. In our case, alarms can be see as the events. Moreover, none of the previous approaches targets Portuguese, so their adaptation would require the development of NLP tools specifically for Portuguese tweets. We should still mention that Portuguese tweets have been exploited for several tasks, such as topic detection [13], sentiment analysis [16], and even for hazard management, such as predicting flu incidence [15].

3   Datasets Explored

Our main goal is to use Twitter to semantically-enrich alarms triggered in the telecommunications network of Portugal Telecom (PT). Our hypothesis is that Twitter users would tweet about the possible causes of the alarm, potentially useful for network management. Alternatively, customers would complain about their quality of service (e.g. slow/unavailable network). Before starting to develop, we enrolled on an exploratory work, in order to become more familiar with the domain of our problem and make a preliminary assessment on the suitability of Twitter for this purpose. For such, we explored two datasets: one with all the alarms triggered during a month, and, for the same period, all the tweets written in Portuguese and published in Portugal (within API limits). This section describes both datasets, followed by information on their combination.

3.1 Description

The alarm dataset, provided by PT's Alarmistics team, has a total of 873k alarms, triggered between 30th September and 29th October 2014 by the Alarm Manager system. Each alarm is characterised by the following properties: creation time (date), archive time (date), local code (according to a 3-level hierarchy: network group, local network, station area), technology (e.g. 3G, 4G, IPTV), entity and problem. The duration of an alarm is obtained by the difference between the archive and the creation times. The last two properties were obfuscated due to privacy reasons. Moreover, we noticed that alarms in the same place and during an overlapping period of time were common. Therefore, we clustered them, such that alarms in the same cluster would share the local code, entity and problem fields, and would occur in overlapping time periods. After this procedure, we were left with 551,513 alarms.

For the same period of time, we collected 498,896 tweets, which, according to Twitter, were written in Portuguese and published in Portugal. This was done through Twitter's public Streaming API[2], a service that provides a continuous stream of tweets, in real-time, corresponding to a random sample of all public statuses, estimated to 1 % of all the published tweets. Each tweet is characterised by the following relevant properties, among others: ID, timestamp, location, coordinates, text, language.

To enable queries to the datasets, a relational database was created with a table for the alarms and another for the tweets, both populated with their data.

3.2 Pairing Alarms and Tweets

In order to match tweets and potentially related alarms, each alarm was paired with tweets published at the same time and location. As tweets have a location property and location names can be extracted from the alarms local code, in the presented experiments, we did not use the tweets coordinates We observed a huge dispersion in the duration of an alarm, on average, 74 min, with a standard deviation of 411 min. An alarm may last for only a few minutes, but can also take several days. For instance, 188,082 alarms lasted for less than 5 min, and 353,867 for less than 10. At the other end, 5,708 alarms lasted for more than one day, and 158 for more than one week. Since Twitter users might notice the alarm cause before it is triggered, and they might keep talking about it after the alarm has been archived, for each alarm, we also paired tweets published between 15 min before the alarm creation and 60 min after archive time. Also, since an

---

[2] https://dev.twitter.com/streaming/public.

alarm could lead to denial of service, this has in mind that affected users may only be able to use the Internet when the problem is solved.

Only tweets that matched either the second (local network) or the third level (station area) of the alarms location hierarchy were considered. These are typically the name of a city and the parish or city area. The top level (network group) is typically the name of a district and was considered to be too general. Figure 1 shows an alarm and some tweets matched this way.

Created	Archived	...	Technology	LocalNetwork	StationArea
2014-10-10 00:31:59	2014-10-11 13:09:22	...	TEC_X	LOC_Y	LOC_Z

TwitterID	Timestamp	Location	Text
520662632240783361	2014-10-11 13:09:22	LOC_Z	*Adoro quando me ignoram*
520680458729029635	2014-10-10 20:50:19	LOC_Z	*Será que se eu ligar a televisão consigo ver?*
520712340216758272	2014-10-10 22:01:09	LOC_Z	*Vou tomar banho ??*

Fig. 1. Alarm and tweets published from the alarm's location, while it was on. Real alarm location name and technology are not provided due to confidential reasons.

After this, 236,227 tweets were matched with alarms, based on their timestamp and location name, which is about a 47 % of all the tweets in the dataset. This is a little less surprising if we add that the 67,605 alarms from the clustered dataset with at least one matching tweet lasted, on average, for about 28 h, with a standard deviation of 78 h. Most of the alarms without matching tweets lasted for only a few minutes and were triggered for small locations.

As relevant tweets may be published from different locations than the alarm, we also matched those published at the same time and mentioning the alarm location. We did not use a named entity recogniser because we are not aware of a such a system, available, and trained with Portuguese tweets. Given the specificities of this kind of text, where location names and other entities are frequently uncapitalized or abbreviated, we would get a noisy analysis either way, and chose the simplest approach for exploration. Only 7,891 tweets were matched this way, and most of the results using this subset are not significant.

4   Experimentation

This section reports on experiments performed to assess the suitability of Twitter for semantically-enriching the telecommunication alarms. They should be seen as exploratory experiments that aim to provide useful insights for the future of this project and for other researchers willing to exploit Twitter.

4.1   Manual Labelling of Tweets

To have some clues on the kind of tweets we could expect during an alarm, we generated two random samples of 200 tweets matching the time of an alarm and: (a) one matching also the location; (b) another mentioning the name of the location. For each sample, we manually tagged the tweets as follows: (a) mentions an occurrence that could be the cause of the alarm; (b) complaints on the service quality; (c) not relevant for our task.

The majority of the tweets were not relevant at all for this task. None of the samples contained a potential complaint. The first sample contained five tweets mentioning a meteorological event, such as rain, and the second contained six mentioning meteorological events or loss of electricity. Figures 2 and 3 display those tweets and their rough translation. We further confirmed that the second sample was noisy and some location names were matched by accident. There are locations in Portugal named *luz* (light/electricity) and *guia* (drive), in Fig. 3, among others, such as *tomar* (to take).

esta a chover, outra vez?
(it is raining, again?)
ta a trovejar bue, vou morrer aqui
(there is much thunder going on, I'm going to die here)
Ta um relâmpago
(there is a lightning)
Mandei um sms a minha mãe que esta no quarto ao lado a dizer que esta a chover
(I sent a sms to my mother who is in next door's bedroom saying it is raining)
hoje apanhei uma chuva do crl
(today I caught heavy rain)

Fig. 2. Tweets published in the same location and at the same time as an alarm, mentioning a meteorological event.

4.2 Keyword Search

For additional insights on the kind of tweets and possible relations to the alarms, we queried our database for tweets with specific keywords, which could be somehow related to potentially problematic events. Those included words related to:

- Bad weather: *chuva* (rain), *chover* (to rain), *trovoada* (thunder), *cheia* (flood), *vento* (wind), ...;
- Networks: *net* (short for network), *tv* (tv), *telefone* (phone), *telemóvel* (mobile phone), *meo* (short for PT's network), *luz* (electricity), ...
- Generic problems: *problema* (problem), *acidente* (accident), ...
- Swearing: *merda* (shit), *bosta* (shit), ...

For each word, we compared the frequency of their mentions in tweets matching the alarms location and time, in opposition to tweets not matched with any alarm. Table 1 has a selection of those numbers, together with the results of a Z test – statistically significant at the 95 % confidence level if $Z > 1.96$. This showed that, although the proportion of several words is slightly higher during the alarms, only the differences of the *rain*-related and the *swearing* words are statistically significant. We may thus speculate that there are more alarms when it is raining, and that people swear more during an alarm, possibly complaining about their quality of service. We cannot draw additional conclusions.

*E o sol apareceu! #greenfest #figodaindia #icecream #sol FIARTIL (Feira de Arte-sanato do **Estoril**) http://t.co/M9K8xGo8TB*
(An the sun came out! #greenfest #figodaindia #icecream #sol FIARTIL (Feira de Artesanato do Estoril) http://t.co/M9K8xGo8TB)
O tempo está a decidir se chove ou se faz sol ???????? **Sobral de Monte Agraço** *http://t.co/RZlMUZsayi*
(The weather is deciding whether it rains or it is sunny ????????   Sobral de Monte Agraço http://t.co/RZlMUZsayi)
*Em **Oeiras** está trovoada, a vida é bela*
(In Oeiras there is thunder, life is beauty)
***Guia** até à casa da Cheila, tive lá um bocado e voltei para casa, nem deu para sair porque chuva*
(Drive to Cheila's, I was there for a bit and came back home, it wasn't even possible to leave because it was raining)
*Eh pah! Ja e a 2da vez que falha a **luz** esta semana...*
(Sheesh! It is the second time electricity fails this weak...)
*Sou so eu que nao tenho **luz**?*
(Is it just me who has no electricity?)

Fig. 3. Tweets published at the same time as an alarm, mentioning both the location of the alarm and a possible cause of the alarm. Location names are in bold.

4.3 Case Study: Complaining About the Network Service

Although the difference was not significant, we manually labelled all the tweets published during an alarm, and mentioning the word 'net', to check how many were possible alarm causes and/or complaints. Since this word denotes only the data network, we did not consider tweets matched to an alarm on the ip television network (IPTV). From the 301 tweets labelled, 132 ($\approx$43 %) were possible complaints. Figure 4 shows some of them, starting with the only two that additionally mention potential alarm causes: weather and electricity loss.

To check whether complaining in Twitter was a common practice, we selected a random sample of 100 that did not match any alarm and contained the word 'net'. This time, the percentage of complains was 31 % which is still high, but significantly lower than 43 %. Here, we should recall that not all Twitter users are PT customers, and their network providers might have problems at different times. The name of the network operator is hard to identify because it is rarely mentioned in the tweet – notice Table 1, which shows only 114 mentions of 'meo', a short name for PT, in the full set tweets. In addition to this, there might be problems that affect only the personal network of a specific customer and, consequently, do not trigger any alarm.

4.4 Event Classification in Tweets

In a final experiment, we were inspired by some of the works referred in Sect. 2 on event detection from Twitter. We used a set of text classifiers, developed in a previous project, for identifying types of events mentioned in tweets. Twelve classifiers

Table 1. Some keywords and their occurrences in tweets matched with alarms by the local name vs their occurrences in the full set of tweets.

Words	Tweets vs Alarms				Z Significance	
	Match		No match			
	#	Prob	#	Prob	Score	@95 %
chuva	447	0.19 %	280	0.11 %	7.6390	Yes
bosta	65	0.03 %	67	0.03 %	0.4355	No
chover	335	0.14 %	219	0.08 %	6.1880	Yes
trovoada	123	0.05 %	70	0.03 %	4.5586	Yes
cheia	898	0.38 %	934	0.36 %	1.4321	No
vento	56	0.02 %	49	0.12 %	1.2281	No
luz	151	0.06 %	150	0.06 %	0.9788	Yes
meo	42	0.02 %	72	0.03 %	−2.2473	No
net	525	0.22 %	590	0.22 %	−0.1773	No
acidente	34	0.01 %	23	0.01 %	1.8599	Yes
problema	355	0.15 %	360	0.14 %	1.2328	No
tv	178	0.08 %	255	0.10 %	−2.6023	No
telemóvel	655	0.28 %	743	0.28 %	−0,3729	No
telefone	57	0.02 %	113	0.02 %	1.8117	No
merda	3727	1.58 %	3818	1.45 %	3.5884	Yes
crl	2625	1.11 %	2660	1.01 %	3.3944	Yes
puta	1267	0.54 %	1276	0.49 %	2.5042	Yes
foda	656	0.28 %	627	0.24 %	2.7154	Yes

were used, each learned for the following event types: accident, celebration, ceremony, concert, exhibition, judicial, manifest, meeting, nature, political, show and sports. Each classifier had been trained with 200 Portuguese tweets, half manually labelled as positive and another half as negative, for their event type. All the training tweets were retrieved from Twitter during January and February 2014 and manually selected to fill the 50 % proportion of positive and negative examples for each event type. A single tweet could belong to one, more, or no type. Labelled tweets were imported to the Mallet toolkit[3], which converts input text to features and includes several text classification algorithms out-of-the-box. In this case, the Maximum Entropy algorithm was used, because it lead to the best results in a 10-fold cross validation. For each event type, accuracy ranged from 76 % (show) and 78 % (celebration) to 88 % (sports) and 91 % (concert). Accident and nature were both at 84 %.

Although the classifiers are quite rudimentary — they rely only in Mallet's black-box for identifying the features and learning from them — we used the

---

[3] http://mallet.cs.umass.edu/.

A minha net está como o tempo lá fora. Uma valente merda
(My net is like the weather outside. A real shit)
Fiquei sem luz e sem net assim de repente , q cena do mal
(I was without electricity and net out of a sudden, what a bad thing)
A net em minha casa decidiu não funcionar ????
(The net in my house decided no to work ????)
A net está super lenta
(The net is super slow)
A net da OPERATOR_X tem tado uma bela merda
(OPERATOR_X's net has been a real shit)
Ke nervos.. A net esteve 4 horas sem dar -.-
(So nervous.. The net was 4 hours without working -.-)
@im_a_mermaiid queriia ter vindo mais cedo mas a net num deu a tarde toda :c
(@im_a_mermaiid I wanted to have come earlier but the net did not work the entire afternoon :c)
a minha net é uma merda -.-
(my net is shit -.-)

Fig. 4. Tweets published in the same location and at the same time as an alarm, and containing the word 'net' in a complaint on the quality of their network.

Table 2. Tweets classified automatically according to mentioned event types, during and not during an alarm.

Event	Tweets vs Alarms				Z Significance	
	Match		No match			
	#	Prob	#	Prob	Score	@95%
accident	69	0.03%	73	0.03%	0.2963	No
celebration	332	0.14%	361	0.14%	0.2942	No
ceremony	163	0.07%	187	0.07%	−0.2918	No
concert	413	0.17%	515	0.20%	−1.7378	No
nature	332	0.14%	302	0.11%	2.5311	Yes
show	92	0.04%	117	0.04%	−0.9646	No
sports	116	0.05%	138	0.05%	−0.5369	No

results of the tweet classification to test whether there was any kind of event more frequently mentioned during an alarm. Similarly to the keywords experiment, this was achieved through querying our relational database. The obtained results for the most relevant event types are presented in Table 2. They show that, except for the *nature* type, all events are mentioned in the same proportion, during and not during an alarm. Events of the type *nature* are more frequently mentioned in tweets matched with alarms. According to a Z-test, the latter result is also statistically significant. This is in agreement with the previous experiments, where we noticed that rain-related words were used more frequently during an alarm.

5   Concluding Remarks

We have reported on an exploratory work towards the utilization of Twitter for semantically-enriching alarms triggered in a Portuguese telecommunications network. Preliminary experiments, presented in this paper, showed that, although not very often, users tweet about the weather and complain about the quality of their network connection. For other events or keywords, we failed to identify strong correlations between tweets and alarms.

The main problem we faced is that useful tweets are always shuffled in a very noisy set, which makes it harder to identify them. This suggests that Twitter is not the most suitable means for achieving our purpose, at least in Portugal. Despite its growing popularity, several reports one can find in the Web show that Twitter has not had as much penetration in Portugal, when compared to other countries[4]. Other facts that contribute to this negative result include the limited number of tweets available through Twitter's public streaming API, which are just a sample of all the tweets, more precisely, an estimated 1 % (and we should not ignore the small population of Portugal, as compared to other countries, such as the United States of America); and the long duration of the matched alarms (on average 28 h), which adds too much noise and increases the number of irrelevant tweets matched.

These experiments were repeated more recently with different criteria for pairing tweets and alarms. Those included different times before creation and after archive (e.g. minimum alarm duration set to 5 s and maximum to 24 h), and for a different time period (March 2015). Even though, due to the better weather, rain-related tweets were much less in March, they were still statistically more frequent during an alarm, which strengthens our main conclusion. On the other hand, assumptions on the complaints were not confirmed.

In the near future, we will start to develop a prototype for our purpose, which should involve training several new automatic classifiers, for which we should consider using an alternative to Mallet, in order to have more control over the used features. The set of event types should also be re-arranged and new training datasets will be selected. And it would definitely be useful to train an additional classifier for complaining tweets, in order to help PT measuring the impact of their alarms in their customers satisfaction.

To come up with better conclusions, we will also repeat some of the previous experiments, with focus both on the geographic information and still on the time periods. On the former, we will try to match tweets according to their coordinates, instead of just the location name. On the latter, we will test more aggressive time windows. For instance, we might ignore alarms with less than a few minutes, as they should have lower impact, or we might only pair tweets

---

[4] Although we could not find a specific study on the usage of Twitter in Portugal, our country is never listed in the top countries in terms of percentage of Twitter users. Also, in the World Map in http://www.beevolve.com/twitter-statistics#b1 (retrieved on March 2015) Portugal had one of the lightest shades of blue, which corresponds to the countries with less Twitter users.

posted until 15 min after the alarm was triggered, to check if people tweet about the event that triggering event. We might as well consider the technology that triggered the alarm (mobile network or television).

Finally, to minimize the strong limitations of Twitter, additional information sources will also be exploited. These should include news websites and others that provide structured information on weather alerts and cultural events. When compared to Twitter, news text should be less challenging to process with traditional NLP tools. On the other hand, we will have to deal with certain limitations. For instance, the flow of news is not as fast as tweets, and there are many news reporting to events occurring in the previous hours our days.

Acknowledgements. This work was developed in the scope of a project funded by Portugal Telecom Inovação e Sistemas, under the cooperation and innovation programme between PT and academic organisations.

References

1. Alonso, O., Shiells, K.: Timelines as summaries of popular scheduled events. In: Proceedings of 22nd International Conference on World Wide Web Conference, Companion, WWW 2013, pp. 1037–1044. WWW/ACM, Geneva (2013)
2. Avvenuti, M., Cresci, S., Marchetti, A., Meletti, C., Tesconi, M.: Ears (earthquake alert and report system): a real time decision support system for earthquake crisis management. In: Proceedings of 20th ACM SIGKDD International Conference on Knowledge Discovery and Data Mining, KDD 2014 pp. 1749–1758. ACM, New York (2014)
3. Becker, H., Naaman, M., Gravano, L.: Beyond trending topics: real-world event identification on Twitter. In: Proceedings of 5th International Conference on Weblogs and Social Media, ICWSM 2011. AAAI Press (2011)
4. Chua, F.C.T., Asur, S.: Automatic summarization of events from social media. In: Proceedings of 7th International Conference on Weblogs and Social Media, ICWSM 2013 (2013)
5. Culotta, A.: Lightweight methods to estimate influenza rates and alcohol sales volume from Twitter messages. Lang. Resour. Eval. **47**(1), 217–238 (2013)
6. Ghahremanlou, L., Sherchan, W., Thom, J.A.: Geotagging twitter messages in crisis management. Comput. J. **58**(9), 1937–1954 (2015). doi:10.1093/comjnl/bxu034
7. Guy, M., Earle, P., Ostrum, C., Gruchalla, K., Horvath, S.: Integration and dissemination of citizen reported and seismically derived earthquake information via social network technologies. In: Cohen, P.R., Adams, N.M., Berthold, M.R. (eds.) IDA 2010. LNCS, vol. 6065, pp. 42–53. Springer, Heidelberg (2010)
8. Li, C., Weng, J., He, Q., Yao, Y., Datta, A., Sun, A., Lee, B.S.: TwiNER: named entity recognition in targeted Twitter stream. In: Proceedings of 35th International ACM SIGIR Conference on Research and Development in Information Retrieval, SIGIR 2012, pp. 721–730. ACM, New York (2012)
9. Longueville, B.D., Smith, R.S., Luraschi, G.: "OMG, from here, i can see the flames!": a use case of mining location based social networks to acquire spatio-temporal data on forest fires. In: Zhou, X., Xie, X. (eds.) Proceedings of 2009 International Workshop on Location Based Social Networks (GIS-LBSN), pp. 73–80. ACM (2009)

10. Mao, H., Shuai, X., Kapadia, A.: Loose tweets: an analysis of privacy leaks on Twitter. In: Proceedings of 10th Annual ACM Workshop on Privacy in the Electronic Society, WPES 2011, pp. 1–12. ACM (2011)

11. Pak, A., Paroubek, P.: Twitter as a corpus for sentiment analysis and opinion mining. In: Proceedings of 7th International Conference on Language Resources and Evaluation, LREC 2010, ELRA, Valletta, Malta, May 2010

12. Ritter, A., Mausam, Etzioni, O., Clark, S.: Open domain event extraction from Twitter. In: Proceedings of 18th ACM SIGKDD International Conference on Knowledge Discovery and Data Mining, KDD 2012, pp. 1104–1112. ACM (2012)

13. Rosa, H., Carvalho, J.P., Batista, F.: Detecting a tweet's topic within a large number of Portuguese Twitter trends. In: Proceedings of 3rd Symposium on Languages. Applications and Technologies, pp. 185–199. OASICS, Schloss Dagstuhl, June 2014

14. Sakaki, T., Okazaki, M., Matsuo, Y.: Earthquake shakes Twitter users: real-time event detection by social sensors. In: Proceedings of 19th International Conference on World Wide Web, WWW 2010, pp. 851–860. ACM, New York (2010)

15. Santos, J.C., Matos, S.: Predicting flu incidence from Portuguese tweets. In: Proceedings of International Work-Conference on Bioinformatics and Biomedical Engineering, IWBBIO 2013, Copicentro Editorial, pp. 11–18 (2013)

16. Souza, M., Vieira, R.: Sentiment analysis on twitter data for Portuguese language. In: Caseli, H., Villavicencio, A., Teixeira, A., Perdigão, F. (eds.) PROPOR 2012. LNCS, vol. 7243, pp. 241–247. Springer, Heidelberg (2012)

17. Tanev, H., Ehrmann, M., Piskorski, J., Zavarella, V.: Enhancing event descriptions through twitter mining. In: Breslin, J.G., Ellison, N.B., Shanahan, J.G., Tufekci, Z. (eds.) Proceedings of 6th International Conference on Weblogs and Social Media, ICWSM 2012. AAAI Press (2012)

18. Wang, X., Gerber, M.S., Brown, D.E.: Automatic crime prediction using events extracted from Twitter posts. In: Yang, S.J., Greenberg, A.M., Endsley, M. (eds.) SBP 2012. LNCS, vol. 7227, pp. 231–238. Springer, Heidelberg (2012)

19. Wang, Y., Xie, L., Sundaram, H.: Social event detection with clustering and filtering. In: Working Notes Proceedings of MediaEval 2011 Workshop, vol. 807. CEUR-WS.org (2011)

20. Zhang, R., Li, W., Gao, D., You, O.: Automatic twitter topic summarization with speech acts. IEEE Trans. Audio Speech Lang. Process. **21**(3), 649–658 (2013)

21. Zielinski, A., Middleton, S.E., Tokarchuk, L.N., Wang, X.: Social media text mining and network analysis for decision support in natural crisis management. In: Proceedings of 10th International Conference on Information Systems for Crisis Response and Management, ISCRAM 2013, Karlsruher Institut fur Technologie, pp. 840–845 (2013)

Meaning Inference of Abbreviations Appearing in Clinical Studies

Efthymios Chondrogiannis$^{(\boxtimes)}$, Vassiliki Andronikou,
Efstathios Karanastasis, and Theodora Varvarigou

National Technical University of Athens,
9 Heroon Politechniou Str, 15773 Athens, Greece
{chondrog, vandro, ekaranas}@mail.ntua.gr,
dora@telecom.ntua.gr

Abstract. The number of publicly available clinical studies is constantly increasing, formulating a rather promising corpus of documents for clinical research purposes. However, the abbreviations used in these studies pose a serious barrier to any text mining technique. This paper presents a study conducted in the above domain, which used specifically developed tools and mechanisms in order to process a number of randomly selected documents from clinicaltrialsregister.eu. The analysis performed indicated that abbreviations appear at a large scale without their long form (aka expansion). In order to assess the abbreviations' true meaning, it is necessary to utilize the appropriate corpus of documents, apply innovative algorithms and techniques to detect their possible expansions, and accordingly select the appropriate ones. Furthermore, the discrimination power of tokens has a distinctive role in abbreviations construction, and hence, it can facilitate the detection of acronym-type abbreviations. Additionally, the expressions in which abbreviations appear, as well as the preceding or following text are of primary importance for selecting the appropriate meaning.

Keywords: Abbreviations · Expansion · Clinical studies · Semantic analysis · Corpus annotation

1 Introduction

Clinical studies provide the means for bringing new chemical products in the market or collecting new evidence for existing interventions. The studies publicly available by the EU Clinical Trials Register (EUCTR) [3] and ClinicalTrials.gov (CTGV) [5] sites exceed 150 thousand with more than 20 thousand new studies being registered every year. Due to the enormous size of the corpus of studies, which is constantly being increased, it's difficult or even impossible to manually process these documents. Consequently, the application of innovative or state of the art text mining techniques is necessary.

Abbreviations (ABR) comprise an important part of a clinical study. They intend to provide short forms of often long texts (aka expansions) so that authors can efficiently use them in the rest of the document. On the other hand, detecting the expansion (EXP) which

© Springer International Publishing Switzerland 2015
J.-L. Sierra-Rodríguez et al. (Eds.): SLATE 2015, CCIS 563, pp. 38–48, 2015.
DOI: 10.1007/978-3-319-27653-3_4

an abbreviation stands for is essential for its comprehension. In general, ABRs have one or more meanings (aka senses) each of which can be expressed with one or more EXPs. In the context of this work, EXPs that are quite similar with one another, after the application of state of the art string text processing techniques (e.g., Upper Limit of Normal and Upper-Normal Limit), are considered identical. Consequently, there is often one to one correspondence between the EXP and its sense, since the possible EXPs of each ABR are composed of totally different words.

Detecting the meaning of ABRs mentioned in clinical studies is quite difficult for humans and even more so for software agents, since they should be able to cope with cases in which ABRs have been explicitly defined in the document (often following specific patterns, i.e. abbreviation syntactic cues) as well as cases in which their EXP is missing (i.e., not mentioned anywhere in the document). The work presented in this paper focuses on the results of a conducted software-based analysis regarding the correct meaning inference of ABRs contained in clinical studies documents.

The document is structured as follows. Section 2 summarizes related work in the domain of ABR detection and EXP provision. Section 3 briefly describes the study approach, methodology and involved tools. Section 4 analytically presents the main study findings, while Sect. 5 expands this discussion to secondary findings and parameters that could be taken into account and are to be probably covered in future work. Finally, Sect. 6 provides a summary of the work's key points.

2 Related Work

Biomedical ABRs have been extensively studied so far and various algorithms and techniques have been proposed for detecting their EXP. Text alignment approaches make an attempt to match ABRs with their corresponding EXPs based on the characters used, such as the algorithm proposed by Schwartz and Hearst [2]. Park and Byrd [17] have also proposed a rule-based approach for detecting ABR-EXP pairs based on the patterns they belong to. The outcome of text alignment techniques can be further improved if syntactic information is used [15]. Machine Learning approaches have also been used for ABRs recognition, such as the supervised machine learning used for creating an ABRs dictionary from MEDLINE [14].

Statistical approaches (e.g., ADAM [16]) can also be used for abbreviation recognition purposes, on condition that the corresponding EXPs appear frequently enough. Consequently, for providing valuable results, they demand a large number of biomedical articles while they also need an adequate amount of computer resources and time. In practice, they can complement acronym-type ABRs recognition techniques by detecting those ABR-EXP pairs where there is no similarity among the characters being used, as for example the MBA system presented in work [9].

In order to find the meaning of ABRs when their EXP missing, the simplest solution is to assign to every instance of an "unknown" ABR the most commonly used meaning [10]. Alternatively, for abbreviation sense disambiguation a supervised machine learning system can be used, such as the one presented in work [11], using a variety of parameters included but not limited to Mesh terms [6] and Concepts Unique Identifiers [12].

3 Study Methodology and Tools

For the analysis of the ABRs used in clinical studies, 141 documents from EUCTR were randomly selected and accordingly the ABRs' meaning was specified. Specifically, for creating a highly variable corpus of documents which adequately represents the ABRs used in clinical studies, all available documents from the EUCTR web site were downloaded by means of a developed software component and accordingly classified into categories based on the number of different ABRs that they contained. Then one or more documents from each category were randomly selected, taking into account the percentage of documents they represent and the total number of documents to be selected.

In order to precisely determine the meaning of ABRs with the least human effort required, a web application was specifically developed (Fig. 1) which enables users to interact with ABRs (highlighted with green color) and accordingly specify their meaning as well as the specific part(s) of the documents in which the ABRs appear with the meaning provided. In general, ABRs appear with the same sense throughout a document [1]. However, the elements which the meaning provided refers to should be precisely determined, in order to handle such exceptional cases (if any) in which an ABR has different meanings depending on the document section it is being used in. A characteristic example is the roman number IV which may also point to "intravenous" administration of a chemical substance in another sentence. Additionally, in some cases, an ABR may appear in a phrase (e.g. HIV-positive or HIV-negative contain the ABR HIV), and hence, it should be specified that the ABR retains its meaning.

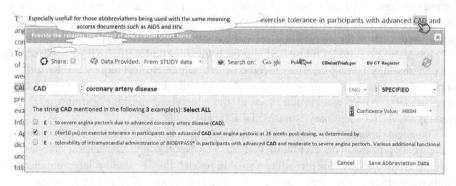

Fig. 1. A screenshot from the application developed for ABR-EXP specification purposes (Color figure online).

In Fig. 1 a screenshot from the web application developed is being presented. The application directly provides the EXP of an ABR when specified in the document using the Schwartz and Hearst algorithm; otherwise, it enables users to search in publicly available sources for the EXP. When all ABRs have been specified, the tool automatically generates an XML document which includes the provided user data and the study details (i.e., ID, Title, URL, etc.) which the data came from.

Throughout the ABRs' definition process, the tool enables users to provide additional data for each ABR-EXP pair, such as whether the EXP has been specified in the document or not and how confident they are for data they provided. The *confidence value* is an essential parameter in order to correctly determine the ABRs' meaning. More precisely, all pairs of ABR-EXP with confidence value medium or low, including those ABRs the meaning of which was unknown to biomedical experts, undergo a review process in which one or more medical or clinical experts are contacted in order for their meaning to be undoubtedly determined.

4   Results Analysis and Main Findings

(a) Expansion Availability. The software-based analysis of data provided indicated that there were on average 13.53 distinct ABRs in every clinical study, from which the 27.25 % had been specified in the document whereas the remaining 72.75 % had been provided without their EXP. From this analysis were deliberately excluded both common Latin ABRs (e.g., i.e., etc.) and Units of measurement (mg, kg, etc.), which often appear without their EXP, as well as the ABRs which appear in the title of each section (e.g., MedDRA) and hence are common to all clinical studies.

Concerning specified ABRs, they often followed their EXP within parentheses or brackets (94 %) or vice versa (4 %), while only 2 % of specified ABRs came from a different syntactic cue. However, it should be noted that in 98 % of the cases where an ABR was enclosed within parentheses or brackets the EXP was to be found in the preceding text, while only in 55 % of the cases where the ABR was followed by text or a phrase enclosed within parenthesis or brackets did that text contain the EXP.

In the case of ABRs provided without their EXP, the whole corpus of clinical studies available at both the EUCTR and CTGV sites was subsequently downloaded (exceeds a total of 200 thousand documents) and further processed by another component. This analysis indicated that 70 % of the user inferred ABR-EXP had been specified in another clinical study. For the remaining 30 % of user inferred pairs, a subsequent analysis indicated that the majority of EXPs could be found in corresponding PubMed [7] articles (i.e., documents in which the same abbreviation appeared in either title or abstract). However, a small number of ABR-EXP pairs was not found since they were clinical trial specific (e.g., LSLV: Last Subject, Last Visit), and hence the corresponding ABRs could either not be found at all or were being used in the PubMed articles with a different meaning.

(b) Pairs Classification. The correspondence between the ABR and EXP characters usually plays an important role for detecting the EXP of acronym-type ABRs. In order to evaluate this, the ABR-EXP pairs specified in all of the above cases were further examined and classified in categories based on the correspondence between ABR characters and EXP tokens (sequence of characters separated by one or more white spaces). The analysis indicated that they can be divided in three categories, with a few examples for each one presented in Table 1.

The first category encompasses those ABR-EXP pairs in which all EXP tokens have contributed in the ABR construction. More precisely, the first character of the EXP's token matches the corresponding ABR characters whereas the rest of ABR characters, if any, do also appear in the EXP tokens with the same order (examples 1–5). The second category encompasses those pairs in which at least one EXP token does not participate (i.e., the first character) in the ABR construction. In general such tokens are function words such as articles and prepositions which enable authors to form grammatically correct human language expressions, but they do not actually add meaning in a phrase or sentence (examples 7 and 8). However, a considerable number of such tokens do not belong in this category (examples 9–12).

Table 1. Classification of ABR-EXP along with an example for each one.

ID	Abbrev.	Expansion (in document)	Description / Comments	
1	CNS	Central Nervous System	The ABR consists of the first character of each EXP token	
2	CrCl	Creatinine Clearance	The ABR consists of the first two characters of each EXP token	
3	CVA	Cerebrovascular Accident	The ABR consists of the first character of each EXP token along with one additional character from the first token	Tightly Linked
4	TZD	Thiazolidinedione	The first ABR-EXP characters matches, while the rest ABR characters appear in the EXP token with the same order	
5	LTP2	Lactate Turnpoint 2	The ABR consists of the first character of the EXP token including the number presented. The ABR character "P" also appears in the 2nd token	
6	SD1	Study Day one	The Arabic number "1" matches with English Word "one" while the rest ABR characters matches with the first character of EXP tokens	
7	ULN	Upper Limit , of Normal	Tokens "of" (stop word) and "," (punctuation) do not contribute in the ABR construction	
8	LVLS	Last Visit of the Last Subject	Tokens "of" and "the" (stop words) do not contribute in the ABR construction	
9	PDE-5	Phosphodiesterase type 5	Token "type" do not contribute in the abbreviation construction	
10	DSM-IV	Diagnostic and Statistical manual of Mental disorders, 4th edition	Tokens "and", "of" (stop words) , "manual, "disorder" and "edition" do not contribute in the abbreviation construction	Loosely Linked
11	ACDA	Acid Citrate Dextrose solution A	Token "solution" do not contribute in the abbreviation construction	
12	CABG	Coronary Artery Bypass Graft procedure	Token "procedure" do not contribute in the abbreviation construction	
13	EKG	Electrocardiogram	"EKG" stands for "Elektrokardiogramm" (German)	
14	SUKL	State Institute for Drug Control	"SUKL" stands for "Státní ústav pro kontrolu léčiv" (Chech)	
15	DL	Dazit	"DL" stands for "Desloratadine", "Dazit" is a Trade Name	Partially / Not Linked
16	AZT	Zidovudine	"AZT" stands for "Azidothymidine", "Zidovudine" is the INN	
17	MDX010	Ipilimumab	"MDX-010" is the product code for "Ipilimumab" (the INN)	
18	C15	Blood and lymphatic diseases	"C15" is the Mesh code for "Hemic and lymphatic diseases"	

The third category includes the pairs in which the ABR has partial or no similarity with its EXP from a characters point of view, since it comes from another EXP rather than the one provided in the document. For instance the EXP provided may be in English whereas the ABR come from another phrase (being conceptually the same) expressed in another language (examples 13 and 14). Also, a chemical product has several names (i.e., chemical name, generic name/s, trade name/s) and hence the ABR may come from another EXP than the one provided in the document (examples 15 and 16). A special case comprises of codes which often are being arbitrarily assigned to concepts when being introduced into a coding system. Such codes have been primarily designed for referencing purposes or communication among software agents. However, they may still be found in documents such as codes from widely used classifications systems (e.g., Mesh – example 18) and especially codes assigned to chemical products before entering in the market (example 17).

Numbers have a distinctive role in the EXP detection process. In general (excluding codes) they appear in both ABR and EXP. However, they may be in different forms, including Roman and Arabic numbers, ordinal numbers, percentages as well as their corresponding phrases in English language (example 6). In example 10 the Roman number is being used in the ABR whereas the corresponding Arabic number in its EXP. However, it should be noted that the sequence of characters "IV" is separated from the rest of the ABR characters. The punctuation characters being used in the ABR (e.g., white spaces or hyphen characters) as well as changes in the characters format (from upper to lower case) or type (from letter to number) may point to *groups of abbreviation characters* that should be examined together.

(c) Tokens Importance. Concerning the pairs of ABR-EXP that belong to the second category, it was observed that an ABR can be "tightly" linked with its EXP if, apart from punctuation characters and function words, one or more tokens are ignored. The analysis of ignored tokens indicated that, in general, they are those words which are not so important in comparison to the other EXP tokens. The *importance* of each token was measured by taking into account the discrimination power of each one across the whole corpus of documents (i.e., studies-count). More precisely, the number of studies each token appeared in was counted (i.e., studies-token-exists) and accordingly its importance calculated, based on the expression (1). The more frequent a token appears the less important it is. In order to overcome the morphological variants of tokens their stem was used based on the Porter stemming algorithm [4]. Also, the importance of tokens was normalized so that it would take values in the range from 0 (not important) to 1 (very important).

$$\text{Token - Importance} = \text{Logarithm - 10 (Studies - Count/Studies - Token - Exists)}$$

$$(1)$$

In Fig. 2, the importance of tokens being ignored (blue) as well as the importance of the remaining EXP tokens (red) is being presented. The 98 % of tokens being ignored have importance in the range [0.0–0.6], with their average importance being close to 0.2. On the other hand, the importance of the remaining EXP tokens covers the whole range. This, in turn, indicates that in the construction of an ABR, in many cases,

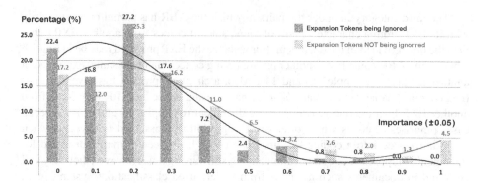

Fig. 2. Classification of the EXP tokens based on their importance.

participate tokens that are not so important or informative. However, in the 77 % of the cases in which an omission was necessary, the average number of the importance of tokens being ignored was lower than the average number of the importance of the remaining EXP tokens, while in almost every case, the most important EXP token had contributed with one or more characters in the construction of the ABR.

The above analysis clearly indicates that authors, when constructing an ABR, tend to ignore the "non-important" tokens. This observation is further supported by the fact that, in 95 % of the cases in which alignment is achieved after ignoring one or more tokens, the number of ABR characters was 3 or more. In other words, if the authors had not omitted such tokens, the length of the ABR would have been more than 4 or 5 characters, which in turn it would have been difficult to remember and use in the rest of the document.

Another important observation is the fact that the importance of tokens ignored is affected by the length of the ABR as well as the corresponding EXP. More precisely, while their length is increased, the average number of ignored-tokens importance is also increased (Fig. 3). Consequently, the authors in order to keep ABR length small, omit words which in some other cases (e.g., when constructing the ABR of a shorter phrase) would have not been ignored.

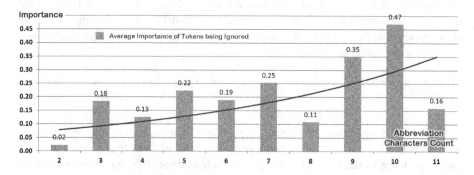

Fig. 3. The average importance of EXP tokens being ignored for each group of ABR.

(d) Abbreviation Expressions. The analysis of documents indicated that ABRs may appear in plural form while they may also participate in one or more expressions. More precisely, an ABR may be used along with other words (e.g., HIV-positive) and/or prefixes (e.g., post-PTA) to form compounds, the meaning of which is affected by the non-ABR components. Based on the examined corpus of documents, 2.1 ABRs on average participate in an ABR expression. Consequently, it's essential to detect the internal ones, which may or may not have been specified in the document. In the latter case, their EXP should be sought in another source, but it's more probable to find the EXP of the internal ABR rather than the whole expression.

Table 2. Dominant ABR Expressions used in Clinical Studies.

Expression	Example(s)	Expression	Example(s)
ABR-positive	HIV-positive	ABR-induced	NRTI-induced
ABR-negative	CRIM-negative	ABR-specific	YMSM-specific
ABR-related	ATGL-related,	ABR-score	FLIE-score, MELD-score
ABR-associated	CVC-associated	anti-ABR	anti-RET, anti-IFN
ABR-based	MR-based	post-ABR	post-PTA, post-CRT
ABR-containing	MPA-containing	pre-ABR	pre-GCRA, pre-TAVR
ABR-like	LDL-like, BMS-like	non-ABR	non-LDL, non-TCC

Table 2 summarizes the "dominant" ABR expressions detected along with a few examples. In fact, the analysis of tokens highlighted this issue, which imposed the analysis of the whole corpus of clinical studies through a semi-automatic process. More precisely, the patterns automatically detected are presented in a descending order based on their frequency of appearance, while for each one of them the specific abbreviation from which they are stemming from was recorded. The combination of two or more ABRs with "and", "or" operators or their corresponding symbols was excluded from the list. Also, the explanation of the ABRs expressions presented was left for our future work.

(e) Expansion Selection. In order to correctly detect the missing EXP of an ABR, the possible EXPs should be found and accordingly the appropriate one be selected. As already mentioned, the 70 % of ABRs inferred by the end user had been specified in another clinical study with the meaning provided. The analysis of the whole corpus of clinical studies indicates that the corresponding abbreviations are highly ambiguous with the average number of possible EXPs being close to 7.0. However, the analysis of the EXPs provided by the end users as well as the possible EXPs detected reveal that in the 87 % of inferred ABR-EXP, the EXP is the *dominant* one, while in almost every case the corresponding EXP has been used in more than one document. It should be noted that the average of EXPs is reduced to 3.1 if the ones being used only in the document provided are ignored.

For improving the percentage of correctly detected ABR-EXP pairs, the broader domain in which each expansion is used was further examined, taking into account the Mesh Terms assigned to each clinical study. More precisely, for each EXP, the corresponding documents were collected and accordingly a graph was created, based on the Tree IDs of Mesh Terms assigned to each one as well as their hierarchy. Then, for each ABR the Tree IDs based on the automatically detected Mesh Terms (e.g., from study title) were retrieved and accordingly the corresponding Tree IDs graphs were examined for finding the best matching EXP. The analysis indicated that only the 85 % of EXPs was correctly detected - a little bit lower than in the previous case.

The terms presented in the close vicinity of each EXP were also examined. Following a similar process, the distinct tokens (i.e., their stem) that appeared in the same sentence with an ABR were gathered, ignoring punctuation characters, stop words and numbers. Then, the preceding and following tokens of the ABRs without EXP were used in the annotated corpus of document for selecting the appropriate EXP. The EXP selected for each ABR was the one for which the sum of the importance (finding c) of tokens matches was the maximum. The outcome of this analysis indicated that the 88 % of ABR-EXP inferred were correctly detected. This approach also proved successful at resolving the meaning of ABR "AR" (a highly ambiguous ABR with up to 10 different EXPs) to "Allergic rhinitis", which was not possible to correctly detect through any of the aforementioned techniques (the dominant EXP for "AR" was "Androgen Receptor").

5   Further Discussion and Future Steps

The ABR-EXP detection process is highly affected by *errors or inconsistencies* which may appear in the documents in both ABR and EXP. For instance, the authors may use an ABR, even in the same document, with a different form than the one specified (e.g., use SDI instead of SD1: Study Day 1). Also, the authors may introduce across the document similar EXPs, which do not accurately match with each other, since they may contain one or more additional punctuation characters (e.g., an additional space or dash) or even have a grammatical error.

Concerning the ABRs mentioned in each document, they were classified in two broad categories; those specified and those inferred. However, in some cases an ABR is *partially defined* in a document, e.g. "HBO2: Hyperbaric O2", where the ABR "O2" stands for oxygen (derived from its molecular formula – not mentioned in the document). Additionally, *nested abbreviations' definitions* should be handled carefully, e.g. "N-methyl-D-aspartate (NMDA) receptor (NMDAR)".

The meaning of ABRs provided without their EXP can be adequately resolved by selecting the dominant EXP for the corresponding ABR. However, the number of correctly detected ABR-EXP pairs can be increased by also taking into account the text that precedes or follows the ABR. The suggestions made can be further improved if not only the tokens presented are taken into account, but also the *semantic class* which they belong to, including their *position* towards abbreviation.

The ABR-annotated corpus of documents, that was an outcome of this work, is available at the following link [13]. However, our intention is the specified ABRs to be further examined and linked with their corresponding terms from widely used nomenclatures, in order for their meaning to be precisely specified (e.g., different semantically equivalent EXPs). Also, the development of a fully-automated system for the analysis of the whole set of clinical studies is planned, with the scope to create a repository of recorded EXPs for all ABRs, including the context that each one is being used within.

6 Conclusion

The performed analysis of the ABR-annotated corpus of clinical studies indicated that ABRs are widely used without their EXP. Specifying the appropriate meaning for each one, presumes the analysis of a larger corpus of documents (not limited to clinical studies) for detecting the possible EXPs and accordingly selecting the appropriate one. Tokens have a significant role in the ABR-EXP detection process, since authors tend to ignore the non-important words, in order to reduce the ABRs' length. Also, the expressions in which an ABR may participate as well as the text that proceeds and follows have a distinctive role for selecting the appropriate meaning.

Acknowledgements. This work is being supported by the OpenScienceLink project [8] and has been partially funded by the European Commission's CIP-PSP under contract number 325101. This paper expresses the opinions of the authors and not necessarily those of the European Commission. The European Commission is not liable for any use that may be made of the information contained in this paper.

References

1. Gale, W.A., Church, K.W., Yarowsky, D.: One sense per discourse. In: Proceedings of the Workshop on Speech and Natural Language HLT 1991, pp. 233–237. New York (1992)
2. Schwartz, S.A., Hearst, A.M.: A Simple algorithm for identifying abbreviation definitions in biomedical text. In: Proccedings of PSB, pp. 451–462 (2003)
3. EU Clinical Trials Register. www.clinicaltrialsregister.eu
4. Porter, M.F.: An algorithm for suffix stripping. Program **40**(3), 211–218 (2006)
5. ClinicalTrials.gov. www.clinicaltrials.gov
6. Medical Subject Headings (MeSH). http://www.nlm.nih.gov/mesh/
7. PubMed. http://www.ncbi.nlm.nih.gov/pubmed
8. Karanastasis, E., Andronikou, V., Chondrogiannis, E., Tsatsaronis, G., Eisinger, D., Petrova, A.: The OpenScienceLink architecture for novel services exploiting open access data in the biomedical domain. In: Proceedings of PCI 2014, pp. 28:1–28:6. ACM, New York (2014)
9. Xu, Y., Wang, Z., Lei, Y., Zhao, Y., Xue, Y.: MBA: a literature mining system for extracting biomedical abbreviations. BMC Bioinform. **10**, 14 (2009)
10. McCarthy, D., Koeling, R., Weeds, J., Carroll, J.: Finding predominant word senses in untagged text. In: Proceedings of ACL 2004, Stroudsburg, PA, USA, pp. 280–287 (2004)
11. Stevenson, M., Guo, Y., Amri, A.A., Gaizauskas, R.: Disambiguation of biomedical abbreviations. In: Proceedings of BioNLP 2009, Boulder, Colorado, USA, pp. 71–79 (2009)
12. McInnes, B.T., Pedersen, T., Carlis, J.: Using UMLS concept unique identifiers (CUIs) for word sense disambiguation in the biomedical domain. In: AMIA 2007, pp. 533–537 (2007)
13. CT abbreviations-annotated corpus. http://147.102.19.246:8080/AbbrAnnotatedCorpus/

14. Chang, J.T., Schütze, H., Altman, R.B.: Creating an online dictionary of abbreviations from MEDLINE. J. Am. Med. Inform. Assoc. **9**(6), 612–620 (2002)
15. Pustejovsky, J., Castaño, J., Cochran, B., Kotecki, M., Morrell, M.: Automatic extraction of acronym-meaning pairs from MEDLINE databases. Stud. Health Tech. I. **84**(1), 371–375 (2001)
16. Zhou, W., Torvik, V.I., Smalheiser, N.R.: ADAM: another database of abbreviations in MEDLINE. Bioinformatics **22**(22), 2813–2818 (2006)
17. Park, Y., Byrd, R.J.: Hybrid text mining for finding abbreviations and their definitions. In: Proceedings of EMNLP 2001 Conference, pp. 126–133 (2001)

Experiments on Enlarging a Lexical Ontology

Alberto Simões[1,2](✉) and José João Almeida[2]

[1] Centro de Estudos Humanísticos, Braga, Portugal
ambs@ilch.uminho.pt
[2] Centro Algoritmi, Universidade do Minho, Braga, Portugal
jj@di.uminho.pt

Abstract. This paper presents two simple experiments performed in order to enlarge the coverage of PULO, a Lexical Ontology, based and aligned with the Princeton WordNet. The first experiment explores the triangulation of the Galician, Catalan and Castillian wordnets, with translation dictionaries from the Apertium project. The second, explores Dicionário-Aberto entries, in order to extract synsets from its definitions. Although similar approaches were already applied for different languages, this document aims at documenting their results for the PULO case.

1 Introduction

Recently, a huge effort has been done to boost the development of wordnet clones for different languages. Portuguese is not an exception. There are different initiatives to create lexical ontologies, linked or not with the original Princeton Word-Net [9] (WordNet.Pr). Examples of such initiatives are Onto.PT [5], PAPEL [6], TeP [8] or Open WordNet-PT [10]. Along with these, another initiative born some months ago: the Portuguese Unified Lexical Ontology (PULO) [12]. It aims at integrating different existing resources into a structure aligned with WordNet.Pr. Recently a joint effort on comparing these projects' history, goals and statuses [4], lead some teams in the direction of cooperation. Nevertheless, each project team continues their own initiatives, enriching and enlarging their resources.

The same happens with PULO. This document describes two experiments performed with the objective of enlarging the number of variants[1]. The kind of experiments are, somehow, similar to some of the previous work, done in order to bootstrap PULO [12] (as we also triangulated three different wordnets, but using probabilistic translation dictionaries), to some of the approaches used to expand GalNet [3], and to create Onto.PT [5]. Although the idea is not new, the thorough description of the process and it's brief evaluation is relevant for future initiatives with other languages.

This short article includes two main sections: Sect. 2 describes the experiment approaches and used resources, while Sect. 3 gives some measures on the quality of the methods application. Finally, it concludes with some brief discussion of the results and future work.

---

[1] This article will use the term *variant* to refer to one of the synonyms of a synset.

© Springer International Publishing Switzerland 2015
J.-L. Sierra-Rodríguez et al. (Eds.): SLATE 2015, CCIS 563, pp. 49–56, 2015.
DOI: 10.1007/978-3-319-27653-3_5

2   Experiments Description

Before running these experiments, PULO included a total of 18.689 variants, distributed by 17.871 synsets (meaning most synsets include only one variant). Table 1 shows how these variants are distributed by morphological category.

Table 1. Distribution of the 18.689 variants prior to the enlargement experiments.

	Nouns	Adjectives	Verbs	Adverbs	Total
Variants	10.421	3.441	4.283	544	18.689

The next subsections describe the two experiments. The first one is based in the triangulation of the Catalan, Galician and Castillian wordnets using translation dictionaries. The second one explores Dicionário-Aberto [11], an open and free definitions dictionary.

2.1   Experiment I: Triangulating Iberian Wordnets

This first experiment uses the wordnets available through Multilingual Central Repository [7], and some translation dictionaries obtained from the Apertium [2] project. Given the reduced number of dictionaries including Portuguese, only the Catalan, Galician and Castillian languages were used. Table 2 shows the sizes for these three wordnets.

Table 2. Summary of sizes for the three used wordnets.

		Nouns	Adjectives	Verbs	Adverbs	Total
Galician	Synsets	18.850	5.092	1.541	349	25.832
	Variants	25.205	8.050	4.145	420	37.820
Catalan	Synsets	36.460	4.148	5.424	1	46.033
	Variants	51.606	7.679	11.577	2	70.864
Castillian	Synsets	26.594	5.180	6.251	677	38.702
	Variants	39.142	6.967	10.829	1.051	57.989

Regarding the translation dictionaries, Table 3 summarizes their sizes. As can be seen, these are quite small dictionaries. This fact was the main reason why the bootstrapping approach [12] used probabilistic translation dictionaries that have a broader coverage. Also, note that most entries in this dictionary have only one translation, reducing the translation ambiguity (which is somewhat desired for a machine translation dictionary, but reduces its applicability for other tasks).

The used algorithm is quite simple. For each synset in the database, that includes at least one variant in any of the three languages, it:

1. Creates a multiset $S_{\mathcal{L}}$ that includes all translations obtained by the translation of all variants for language $\mathcal{L}$. Note that different variants can translate to the same word in Portuguese, so, the multiset tracks the number of times that word was obtained.

Table 3. Translation dictionaries sizes.

Lang. Pair	Nr. Entries	Max Nr. Trans.	Avg. Nr. Trans.
GL–PT	11.003	4	1.07
CA–PT	6.510	7	1.11
ES–PT	12.742	6	1.07

2. Compute the multiset $S = S_{GL} \cup S_{CA} \cup S_{ES}$. This means that, if a Portuguese word was obtained by translating just one variant for each of the source languages, it would have a multiplicity of three. On the other hand, if three variants for just one language generated a Portuguese word, that was not obtained from any of the other languages, its multiplicity would be, as well three. Not giving extra weight if the word was obtained from different languages or every time from the same language was decided in order to keep the algorithm simple.

3. Filter the multiset S, removing all Portuguese variants with a multiplicity of just one. To define this cut line, each variant was checked against current variants in PULO. Figure 1 shows this test. Bars at the left represent variants found in PULO, while bars at the right represent new variants. Given the huge amount of new variants with a multiplicity of 1, it was decided to ignore them (trying to improve accuracy).

4. The bootstrapping approach for PULO used dictionaries obtained from European Portuguese corpora with its old orthography[2]. The dictionaries from Apertium used, essentially, Brazilian orthography that, curiously, is now the correct form for European Portuguese. With that in mind, a simple tool was used to remove variants written in the old orthography, and adding the respective new orthography in case it was not yet present. This process was performed using JSpell morphological analyzer [1].

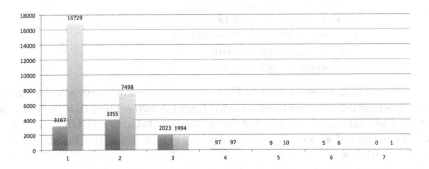

Fig. 1. Number of candidate variants already existing in PULO (left bars) against the new candidates (bars at the right), distributed by their multiplicity in multiset S.

---

[2] Orthography prior to the 1990 agreement, that was officiated in 2008 by the Portuguese Government, and still being, progressively, adopted in Portugal.

This process created a total of 7.229 new variants, and removed 261 of existing variants with the old orthography. Table 4 summarizes the distribution of PULO variants by morphological category after this experiment.

Table 4. Distribution of the 25.657 variants after the first enlargement experiment.

	Nouns	Adjectives	Verbs	Adverbs	Total
Variants	14.062	4.825	6.172	598	25.657

2.2   Experiment II: Synset Extraction from Definitions Dictionary

This second experiment was prepared already with the expectation of a big amount of false positives. Nevertheless, there was interest on confirm that expectation. The main idea was to use Dicionário-Aberto (DA) [11] definitions to construct synsets. DA is partially encoded in TEI[3].

DA definitions are stored in `def` XML elements, with the new line signaling the change of sense[4]. Although XML should ignore spaces and new lines, this decision was taken during the dictionary encoding process for simplicity. Each sense line can include very different types of information. The most common is a standard definition, explaining the concept. In other cases, there are examples, or *see also* references. But there is another kind of definition that is quite interesting for the PULO enlargement process. Some lines include a set of synonyms separated by a semicolon (see an example in Fig. 2). Thus, this second experiment finds lines in DA that are only a sequence of terms separated by a semicolon. For each of these sequences, the list of synonyms, together with the entry head word, are stored.

Exploring the 128.521 entries in DA, 4.842 synsets were found. These synsets have from 3 to 7 synonyms, with an average of 3.14 synonyms per synset. Follow some examples of such synsets:

> acobertar, encobrir, dissimular
> açôfar, pechisbeque, latão
> acordança, melodia, consonância
> acôrdo, convenção, ajuste
> acoroçoado, animado, incitado

In order to map these synsets to PULO synsets, a simple heuristic was used: find an intersection between the synonyms from the two sources that includes, at least, two variants. This means that for a synset obtained from DA $\langle s_1, s_2, s_3 \rangle$, s_i will be suggested as a candidate if there is a synset S in PULO that contains s_j and s_k with $i \neq j \neq k$.

Table 5 show some synsets from PULO (left column) and the aligned synset from DA (right column). In italic are the terms that were used for the alignment.

---

[3] Text Encoding Initiative XML schema, that includes notation to encode different kind of resources from simple books to corpora or dictionaries.

[4] This distinction is, of course, of the responsibility of the original lexicographer.

Abrigo

m.
Resguardo; cobertura.
Protecção.
(Lat. *apricum*)

Fig. 2. Example of an entry from Dicionário Aberto with a line of synonyms.

Table 5. Synsets from PULO at the left, and aligned synset from DA at the right.

cima, cimeira, *cimo*, cumbre, *cume*	vértice, *cimo*, *cume*, culminância
lista, *relação*	tabela, *relação*, catálogo, *lista*
alegria, *prazer*	*prazer*, *alegria*, jovialidade, satisfação, delícia, aprazimento, agrado

This process suggested 1.150 additions. Given this dictionary is quite noisy, and includes a lot of words with old orthography (previous to the 1945 agreement), these suggestions were not added automatically to PULO.

3   Experiments Evaluation

Both evaluations reported here were performed by sampling, given there is no gold standard that can be used to evaluate these candidates, neither the manual power needed to fully (manually) evaluate all candidates from both experiments.

For the second experiment, all suggestions need to be evaluated before being added to PULO. Nevertheless, there was no time to complete that task yet.

3.1   Experiment I

For the first experiment, 200 of the added variants were chosen randomly. This sample included 101 nouns, 39 adjectives, 2 adverbs and 58 verbs.

The evaluation divided these variants into three different categories:

– **Correct Variants:** 152 of the obtained variants were classified as correct. This evaluation was performed looking to the word and the sense gloss. When in doubt, a standard dictionary was used, in order to check if that specific sense was present in the definition.

Follows some examples of variants evaluated in this class, together with its gloss[5]:

  • progredir — get better

---

[5] In these and next examples, the authors decided not to translate the variant itself, as a direct translation will lose part of the cultural/usage meaning.

- corrupção — the state of being corrupt
- aguentar — hang on during a trial of endurance

- **Incorrect Variants:** 40 of the variant candidates were marked as incorrect. Most of these were easy to spot, looking to the synset gloss. Examples of such entries are:
  - pegar — take away to an undisclosed location against their will and usually in order to extract a ransom
  - remeter — make less fast or intense
  - bola — a statement that deviates from or perverts the truth
- **Ambiguous Variants:** There were 8 of the proposed variants that the authors feel they are not incorrect, because there are some situations in which they can be used to represent the synset concept. Nevertheless, as this decision might not be consensual, the variants were classified as ambiguous. Some examples:
  - desnudar — take away possessions from someone
  - puro — spotlessly clean and fresh

Table 6 present these numbers distributed by morphological category, with an accuracy (by sampling) of 76 %[6].

Table 6. Distribution of correct, incorrect and ambiguous variants distributed by morphologic category for first experiment.

	Nouns	Adjectives	Verbs	Adverbs	Total
Correct	82 (81 %)	27 (69 %)	41 (71 %)	2 (100 %)	152 (76 %)
Incorrect	17 (17 %)	9 (23 %)	14 (24 %)	0 (0 %)	40 (20 %)
Ambiguous	2 (2 %)	3 (8 %)	3 (5 %)	0 (0 %)	8 (4 %)
Total	101	39	58	2	200

3.2 Experiment II

Again, for this second experiment, 200 of the candidate variants where chosen randomly and classified in the three classes defined in the previous section. This evaluation resulted in only 115 variant candidates marked for acceptance, while 74 were marked as wrong, and 11 as ambiguous. Table 7 shows the distribution of these candidates by morphologic category. The accuracy[7] on this experiment was 58 %.

Follow some examples of entries obtained throw this experiment for each of the three classes:

- **Correct Variants**
  - constância — persistent determination

---

[6] Given the obtained accuracy and the lack of human resources for a through validation, the authors decided to include the obtained variants without further analysis.

[7] Given the low accuracy and the small number of proposed variants, the authors decided to perform a manual validation prior to their incorporation into PULO.

- sólido — securely in position; not shaky
- truculento — very unpleasant
- **Incorrect Variants**
  - carraceno — very small
  - eduzir — make a subtraction
  - sisudez — a solemn and dignified feeling
- **Ambiguous Variants**
  - bom — to a complete degree or to the full or entire extent
  - aquentar — spur on or encourage especially by cheers and shouts

Table 7. Distribution of correct, incorrect and ambiguous variants distributed by morphologic category for second experiment.

	Nouns	Adjectives	Verbs	Adverbs	Total
Correct	56 (62%)	28 (56%)	31 (52%)	0	115 (58%)
Incorrect	28 (31%)	19 (38%)	27 (45%)	0	74 (37%)
Ambiguous	6 (7%)	3 (6%)	2 (3%)	0	11 (5%)
Total	90	50	60	0	200

4 Conclusions

This article reports two experiments on expanding PULO coverage. Although the used methods are not new, the experiments have shown that these methods can get acceptable accuracy. Even the second method, that used a very noisy and old dictionary (from 1913), could suggest a good set of new variants. Nevertheless, when dealing with semantics, decisions are not consensual, and probably other researchers would accept or reject different number of entries.

Acknowledgements. Thanks to Nuno Carvalho for the proofreading. This work has been partially supported by FCT - Fundação para a Ciência e Tecnologia within the Project Scope UID/CEC/00319/2013.

References

1. Almeida, J.J., Pinto, U.: Jspell - um módulo para análise léxica genérica de linguagem natural. In: Actas do X Encontro da Associação Portuguesa de Linguística. pp. 1–15. Évora 1994 (1995)
2. Forcada, M.L.: Apertium: traducció automàtica de codi obert per a les llengües romàniques. Linguamática **1**(1), 13–23 (2009)
3. Gómez Guinovart, X., Clemente, X.M.G., Pereira, A.G., Lorenzo, V.T.: Galnet: WordNet 3.0 do galego. Linguamática **3**(1), 61–67 (2011)
4. Gonço Oliveira, H., de Paiva, V., Freitas, C., Rademaker, A., Real, L., Simões, A.: As wordnets do Português. In: Simões, A., Barreiro, A., Santos, D., Sousa-Silva, R., Tagnin, S. (eds.) Linguástica, Informática e Tradução: Mundos que se Cruzam, vol. 7, pp. 397–424, March 2015

5. Gonçalo Oliveira, H., Gomes, P.: ECO and Onto.PT: a flexible approach for creating a Portuguese wordnet automatically. Lang. Resour. Eval. J. **48**(2), 373–393 (2014)

6. Oliveira, H.G., Santos, D., Gomes, P., Seco, N.: PAPEL: a dictionary-based lexical ontology for Portuguese. In: Teixeira, A., de Lima, V.L.S., de Oliveira, L.C., Quaresma, P. (eds.) PROPOR 2008. LNCS (LNAI), vol. 5190, pp. 31–40. Springer, Heidelberg (2008)

7. Gonzalez-Agirre, A., Laparra, E., Rigau, G.: Multilingual central repository version 3.0. In: Proceedings of the 8th International Conference on Language Resources and Evaluation (LREC 2012), pp. 2525–2529. ELRA (2012)

8. Maziero, E.G., Pardo, T.A.S., Felippo, A.D., Dias-da-Silva, B.C.: A base de Dados Lexical e a interface web do TeP 2.0. In: VI Workshop em Tecnologia da Informação e da Linguagem Humana, pp. 390–392 (2008)

9. Miller, G.A.: WordNet: a lexical database for English. Commun. ACM **38**, 39–41 (1995)

10. Rademaker, A., Paiva, V.D., de Melo, G., Coelho, L.M.R., Gatti, M.: OpenWordNet-PT: a project report. In: Proceedings of the 7th Global WordNet Conference, pp. 383–390 (2014)

11. Simões, A., Farinha, R.: Dicionário Aberto: um recurso para processamento de linguagem natural. Vice-Versa **16**, 159–171 (2011)

12. Simões, A., Guinovart, X.G.: Bootstrapping a Portuguese WordNet from Galician, Spanish and English wordnets. In: Navarro Mesa, J.L., Ortega, A., Teixeira, A., Hernández Pérez, E., Quintana Morales, P., Ravelo García, A., Guerra Moreno, I., Toledano, D.T. (eds.) IberSPEECH 2014. LNCS, vol. 8854, pp. 239–248. Springer, Heidelberg (2014)

Using Unstructured Profile Information for Gender Classification of Portuguese and English Twitter Users

Marco Vicente[1,2], Joao P. Carvalho[1,3(✉)], and Fernando Batista[1,2]

[1] INESC-ID, Lisboa, Portugal
joao.carvalho@inesc-id.pt
http://www.l2f.inesc-id.pt
[2] ISCTE-IUL - Instituto Universitário de Lisboa, Lisboa, Portugal
[3] Instituto Superior Técnico, Universidade de Lisboa, Lisboa, Portugal

Abstract. This paper reports experiments on automatically detecting the gender of Twitter users, based on unstructured information found on their Twitter profile. A set of features previously proposed is evaluated on two datasets of English and Portuguese users, and their performance is assessed using several supervised and unsupervised approaches, including Naive Bayes variants, Logistic Regression, Support Vector Machines, Fuzzy c-Means clustering, and k-means. Results show that features perform well in both languages separately, but even best results were achieved when combining both languages. Supervised approaches reached 97.9 % accuracy, but Fuzzy c-Means also proved suitable for this task achieving 96.4 % accuracy.

Keywords: Twitter users · Gender detection · Fuzzy c-Means · Supervised methods · Unsupervised methods

1 Introduction

The growth of social networks has produced massive amounts of data. This user-generated information provides clues about users' opinions, daily routines, reaction to events, among other. Twitter, with about 500 million user-generated tweets per day, provides an opportunity for social networking studies [4], and has become the subject of studies seeking to understand public opinion [7]. Unlike other social networks, a user name is the only required field when creating a Twitter profile. There are not even specific fields to indicate information such as gender or age. Nevertheless, the user profile includes optional text attributes that can be used. Previous studies support the hypothesis that users tend to choose real names more often than other forms [2] and, in fact, gender information is most of the times provided either wittingly or unwittingly, for example, in the *screen name* (e.g. "johndoe95" or "marianacruz") or in the *user name* (e.g. "John Doe the best :)" or "the macho man!!!").

© Springer International Publishing Switzerland 2015
J.-L. Sierra-Rodríguez et al. (Eds.): SLATE 2015, CCIS 563, pp. 57–64, 2015.
DOI: 10.1007/978-3-319-27653-3_6

The natural language processing (NLP) problem of gender detection, i.e., deciding if the author of a text is male or female, has been previously applied to Twitter. There are basically two major ways of addressing the problem of gender detection in Twitter: (1) by looking for naming hints included in the unstructured textual profile information; (2) by analyzing the tweet contents. The first approach is *a priori* simpler, but it is highly dependent on the fact that the user must somehow hint its real name in the user name or screen name fields. On the other hand, a single tweet is enough to perform a user's gender detection. The second approach does not need such information since it looks for gender specific information (unwillingly) provided by a user when tweeting. However, it needs each user past tweeting history, and can only give good results for users that tweet a lot and produce enough text. Rao et al. [17] examined Tweets written in English, using Support Vector Machines with character n-gram-features and sociolinguistic features like emoticons use or alphabetic character repetitions. They reported an accuracy of 72.3 % when combining n-gram-features with sociolinguistic features. The state-of-the-art study reported by Burger et al. [6] uses a large multilingual corpora, including approximately 184 k users labelled with gender, 3.3 million tweets for training, and 418 k tweets for testing. They used SVMs, Naive Bayes and Balanced Winnow2 with word and character n-grams as features. Using tweet texts alone they achieved the accuracy of 75.5 %. When combining tweet texts with profile information (*description, user name* and *screen name*), they achieved 92 % of accuracy. A study on Dutch users, using tokens and character n-grams, is reported by Halteren et al. [10]. Only users with significant portions of produced tweets were studied, but using SVMs and token unigrams the study reports 95.5 % accuracy. In this work we try to improve automatic user gender detection in Twitter using the unstructured information found on that user profile.

Using names to detect a user gender is, a priori, a rather trivial task. All that is needed is a good dictionary of names and the will of a user to somehow provide his/her name in the profile. E.g.: the user whose user name is John Gaines, should be male. If the names appearing on the profile are not proper, e.g.: John75, JooohnGaines, or J0hn G4ines, then it is possible to recover the user name (in this case, John) using some simple text/NLP techniques. The problem is that by using such techniques, lots of noisy information might arise. In the previous example, form "John Gaines" we would obtain "John", "Aine" and "Ines". Since both Aine and Ines are female names, we would obtain a conflicting gender info. Nevertheless, using a dictionary of names and basic NLP process, the achieved accuracy is almost 89 % when any form of a name is detected within the "User name" or the "Screen name" fields. It is our contention that this number can be improved by using additional features extracted from such fields.

This paper describes a set of features for gender classification proposed in our previous study [18], which rely on the user's profile unstructured textual information. The main contributions are two-fold: Firstly, we assess the performance of the features using several supervised and unsupervised methods for a Portuguese dataset, in addition to the English dataset used in our previous study. Secondly, we show that the proposed features are compatible with both languages, and that results are improved when merging both datasets. We notice that using unsupervised methods, the increasing amount of data has positive impact on the results. The features can be used to extend gender labelled datasets for researchers.

The paper is organized as follows: Sect. 2 characterizes the data, describes the proposed features and describes our golden set of manually labelled data. Section 3 describes experiments and reports the corresponding results. Section 4 presents the conclusions and prospects about the future work.

2   Data and Features

Experiments performed in this paper use an English and a Portuguese dataset of Twitter users. The English dataset was extracted from one month of tweets collected during December 2014, using the Twitter *streaming/sample* API. The data has been restricted to English geolocated tweets, either from the United States or from the United Kingdom, totaling 296506 unique users. The Portuguese dataset is a subset of the data described in Brogueira et al. [5], and corresponds to a database of Portuguese users, restricted by users that have tweeted during October of 2014 in Portuguese language, and geolocated in the Portuguese mainland.

2.1   Names Dictionaries

In order to automatically associate names that can be found in the user's profile with the corresponding gender, we have compiled a dictionary of English names and a dictionary of Portuguese names. Both dictionaries contain *gender* and *number of occurrences* for each of the names, and focus on names that are exclusively male or female, since unisex names can be classified as male or female. The English names dictionary contains about 8444 names. It was compiled using the list of the most used baby names from the United States Social Security Administration. The dictionary is currently composed of 3304 male names and 5140 female names. The Portuguese names dictionary contains 1659 names, extracted from Baptista et al. [1]. Their work is based on the extraction of names both from official institution lists and from previous corpora. The dictionary is currently composed of 875 male names and 784 female names.

2.2   Feature Extraction

Our experiments use the features proposed in a previous work [18], which are extracted with the dictionaries of names described previously. The profile information is normalized for repeated vowels (e.g.: "eriiiiiiiiic" → "eric") and "leet speak" [9] (e.g.: "3ric" → "eric"). After finding one or more names in the *user name* or *screen name*, we extract the applicable features from each name by evaluating elements, such as "case", "boundaries", "separation" and "position". Each feature has a minimum size threshold (i.e.: the size of the name must have

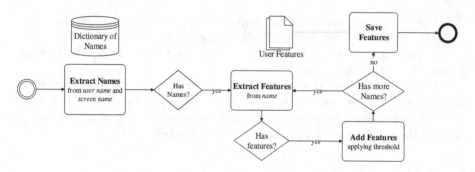

Fig. 1. Feature extraction diagram.

at least a number of characters). Weak features have higher thresholds. If the length of the extracted name is smaller than the threshold, the feature is discarded. The final model uses 192 features. Each element increases the feature granularity. Figure 1 illustrates the feature extraction process.

Consider the screen name "john_gaines" as an example. Three names are present in the dictionary of names and are extracted: "john", "aine" and "ines". The name "aine" has no valid boundaries, since is preceded and succeeded by alphabetic characters. The feature found is weak and the size of the name is lower than the previously defined threshold. Consequently, the name is discarded. The name "ines" has a valid end boundary, as it is not succeeded by alphabetic characters. The feature for a name with correct end boundary has a threshold of 5 and the name is discarded (e.g.: in the case of the screen name "kingjames", the name "james" would not be discarded). Finally, the name "john" has a valid end boundary and starts at the beginning of the screen name. The feature for names with this boundary (valid end boundary) and this position (start of screen name) is 3. The name "john" is selected along with its features.

About 243522 English users (82 %) and 15828 Portuguese users (58 %) trigger at least one gender feature.

2.3 Labelled Data

In order to perform the evaluation, we manually labelled a randomly selection of Portuguese users with gender information and used the existing labelled English dataset [18]. The corresponding gender was assigned by manually analyzing and validating users based on their user name/screen name, their profile picture and checking if associated blogging websites corresponded in gender. All users in our labelled datasets contain at least a sequence that matches a name in our dictionary of names. The English labelled dataset has 748 users: 330 male users and 418 female users. The Portuguese labelled dataset has 716 users: 249 male users and 467 female users. The majority of the users are female, which is consistent with the work of Heil et al. [11] that performed a study of correlation between name and gender, and estimates that 55 % of Twitter users are female.

Table 1. Features extracted from each profile and their properties.

	English		Portuguese	
	User name	Screen name	User name	Screen name
Number of extracted features	3221	1925	1798	2404
Leet related features	291	208	17	15
Repeated vowels related features	20	48	4	122
Average name length (chars)	5.4	5.3	5.2	4.7
Percentage of rejected names	29 %	73 %	13 %	16 %

Table 1 shows the number of features that can be extracted from the manually labelled subset as well as statistics for the extracted names in each one of the profile attributes. For English we observe more occurrences of features in user names (63 % against 37 % in screen names). The frequency of "Leet speak" is consistent with the general features distribution. As expected, repeated vowels occur more in *screen names* because they must be unique for all Twitter users, unlike *user names* that impose no restrictions to their content. For Portuguese we observe more occurrences of features in *screen names* (57 % vs 47 % in user names). Repeated vowels related features occur more frequently in Portuguese *screen names*. The English data reveals that names in *screen name* are more unreliable. That is due to the *screen name* being a unique string without spaces, which leads to a higher uncertainty when extracting possible names. Names in *screen name* are more unreliable in English users than in Portuguese users (73 % versus 16 %). A similar discrepancy can be found in Chen et al. [8], that achieved better results with Portuguese and French than with English and German, when identifying language origin of names using trigrams of letters.

3   Experiments and Results

This section describes the results obtained on the English and Portuguese datasets, and the dataset containing both English and Portuguese users, when applying supervised and unsupervised approaches based on the proposed features. The supervised methods include: Multinomial Naive Bayes (MNB) [15], a variant of Naive Bayes, Logistic Regression [13], and Support Vector Machines (SVM) [12,16]. The unsupervised methods include Fuzzy c-Means clustering (FCM) [3] and k-means [14]. The fuzzy logic module from the scikit-learn toolkit[1] was used for implementing FCM, and all the other methods were applied through Weka[2], a collection of open source machine learning algorithms and a collection of tools for data pre-processing and visualization.

While the supervised based methods use labelled data to build a model, that is not the case of unsupervised methods, which group unlabelled data into clusters.

---

[1] https://github.com/scikit-fuzzy/scikit-fuzzy.

[2] Weka version 3-6-8. http://www.cs.waikato.ac.nz/ml/weka.

Table 2. Gender classification results for supervised and unsupervised methods.

	English		Portuguese		English + Portuguese	
	Accuracy	kappa	Accuracy	kappa	Accuracy	kappa
Logistic regression	93.7 %	0.87	97.6 %	0.95	96.3 %	0.92
Multinomial naive bayes	**97.2 %**	**0.94**	**98.3 %**	**0.96**	**97.9 %**	**0.96**
Support vector machines	96.4 %	0.93	97.8 %	0.95	97.4 %	0.95
k Means clustering	67.3 %		70.1 %		67.8 %	
Fuzzy c-Means	**96.0 %**		94.4 %		**96.4 %**	

For that reason, we will first describe experiments using labelled data only, and then will extend the analysis to all the data, but restricting the experiments to unsupervised methods only. Experiments using supervised methods use the labelled data for training and used a 5-fold cross-validation. Experiments using unsupervised methods use all data for creating two different clusters, the labelled data was used for validation, and each cluster was assigned to the class with more elements from that cluster. In terms of setup, k-means was set to use the Euclidean distance, centroids are computed as a mean, and the seed was set to 10. In order to use the FCM clustering algorithm, the data has been converted into a matrix of binary values, and we have used 1000 iterations, and the Euclidean distance. All experiments consider binary features.

Results achieved with each one of the methods are summarized in Table 2. The first 3 rows show the performance for supervised methods. Results from the last two columns were achieved by combining both the English and the Portuguese labelled subsets. MNB achieved the best performance for both languages, and achieves even better performance for the merged subset of users, achieving about 98 % accuracy, proving that datasets can be combined and that features are compatible with the two languages. The achieved performance suggests that the proposed features can be suitable to discriminate the user's gender for both languages. The last two rows of the table summarizes the performance for unsupervised methods. FCM obtains the correct gender for about 96.0 % of the English users and about 94.4 % of the Portuguese users when all the data is used. k-means achieves a much lower performance for both languages. The last column of the table shows the results when English and Portuguese data are combined. With such dataset, FCM achieves the best results so far, outperforming individual results obtained for each language.

Our proposed features compare well with the performance achieved by other state-of-the art research, despite being applied to only about 82 % of English users. For example, Burger et al. [6] uses the winnow algorithm with n-grams extracted from the user's full name and obtain 89.1 % accuracy for gender detection.

We have performed additional experiments in order to assess the impact of using increasing amounts of data. Figure 2 shows the impact of the amount of data on the performance of FCM, revealing that it has positive impact until reaching the 50 k users. Above that threshold, the accuracy tends to remain stable, which may be due to our relatively restricted set of users.

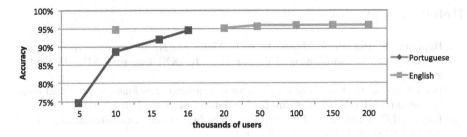

Fig. 2. Impact of the amount of data on the performance, for Portuguese and English.

4   Conclusions and Future Work

We have described an approach to automatically detect the gender of Twitter users, using unstructured profile information. A number of name related features is evaluated on a dataset of about 244 K English users and a dataset of about 16 k Portuguese users. Different supervised and unsupervised approaches are used to assess the performance of the proposed features. The proposed features proved to be good for discriminating the user's gender in Twitter, achieving about 97.9 % accuracy using a supervised approaches, and about 96.4 % accuracy using the unsupervised approach based on Fuzzy c-Means, which also proved to be very suitable for this task. Our features proved to be compatible between the English and Portuguese datasets of Twitter users. Experiments show that by combining datasets of English and Portuguese users, the performance can be further increased. The performance of Fuzzy c-Means significantly increased when more data was used for learning the clusters. Above 50 k users, the performance stabilizes, probably to the relatively small amount of labelled data. Fuzzy c-means proved to be an excellent choice for the gender detection on Twitter since: (i) it does not require labelled data, which is relevant when dealing with Twitter; (ii) its performance increases as more data is provided; and (iii) it achieves a performance almost similar (1.5 % lower) to the best supervised method.

Future work will encompass the creation of an extended labelled dataset in a semi-automatic fashion, based on an automatic annotation provided by our proposed features. Using such labelled dataset, we will associate the textual content provided by the users with their gender and create gender models, purely based on the text contents. In addiction, we will create age models for our Twitter dataset.

Acknowledgements. This work was supported by national funds through Fundação para a Ciência e a Tecnologia (FCT) under project PTDC/IVC-ESCT/4919/2012 and funds with reference UID/CEC/50021/2013.

References

1. Baptista, J., Batista, F., Mamede, N.J., Mota, C.: Npro: um novo recurso para o processamento computacional do portugus. In: XXI Encontro APL, December 2005
2. Bechar-Israeli, H.: From <bonehead> to <clonehead>: nicknames, play, and identity on internet relay chat. Comput.-Mediated Commun. 1(2) (1995)
3. Bezdek, J.C., Ehrlich, R., Full, W.: Fcm: The fuzzy C-means clustering algorithm. Comput. Geosci. 10(23), 191–203 (1984)
4. Brogueira, G., Batista, F., Carvalho, J.P., Moniz, H.: Portuguese geolocated tweets: an overview. In: Proceedings of the International Conference on Information Systems and Design of Communication, ISDOC 2014, pp. 178–179. ACM, New York (2014). http://doi.acm.org/10.1145/2618168.2618200
5. Brogueira, G., Batista, F., Carvalho, J.P., Moniz, H.: Expanding a database of Portuguese tweets. In: 3rd Symposium on Languages, Applications and Technologies SLATE 2014. OpenAccess Series in Informatics (OASIcs), vol. 38, pp. 275–282 (2014)
6. Burger, J.D., Henderson, J., Kim, G., Zarrella, G.: Discriminating gender on twitter. In: EMNLP 2011, pp. 1301–1309. ACL (2011)
7. Carvalho, J.P., Pedro, V., Batista, F.: Towards intelligent mining of public social networks' influence in society. In: IFSA World Congress and NAFIPS Annual Meeting (IFSA/NAFIPS), pp. 478–483, Edmonton, Canada, June 2013
8. Chen, Y., You, J., Chu, M., Zhao, Y., Wang, J.: Identifying language origin of person names with n-grams of different units. In: IEEE ICASSP 2006, vol. 1, p. I, May 2006
9. Corney, M.W.: Analysing e-mail text authorship for forensic purposes. Ph.D. thesis, Queensland University of Technology (2003)
10. van Halteren, H., Speerstra, N.: Gender recognition on dutch tweets. Comput. Linguist. Neth. J. 4, 171–190 (2014)
11. Heil, B., Piskorski, M.: New twitter research: men follow men and nobody tweets. Harvard Bus. Rev. 1, 2009 (2009)
12. Keerthi, S., Shevade, S., Bhattacharyya, C., Murthy, K.: Improvements to platt's SMO algorithm for SVM classifier design. Neural Comput. 13(3), 637–649 (2001)
13. Le Cessie, S., Van Houwelingen, J.C.: Ridge estimators in logistic regression. Appl. Stat. 41(1), 191–201 (1992)
14. MacQueen, J.: Some methods for classification and analysis of multivariate observations (1967). http://projecteuclid.org/euclid.bsmsp/1200512992
15. McCallum, A., Nigam, K., et al.: A comparison of event models for naive bayes text classification. In: AAAI-98 Workshop on Learning for Text Categorization, vol. 752, pp. 41–48 (1998)
16. Platt, J., et al.: Fast training of support vector machines using sequential minimal optimization. In: Advances in Kernel Methods—Supports Vector Learning 3 (1999)
17. Rao, D., Yarowsky, D., Shreevats, A., Gupta, M.: Classifying latent user attributes in twitter. In: Proceedings of the 2nd International Workshop on Search and Mining User-Generated Contents, pp. 37–44. ACM (2010)
18. Vicente, M., Batista, F., Carvalho, J.P.: Twitter gender classification using user unstructured information. In: FUZZ-IEEE 2015, IEEE International Conference on Fuzzy Systems. IEEE Xplorer, Istanbul, Turkey (Accepted)

Yet Another Suite of Multilingual NLP Tools

Marcos Garcia[✉] and Pablo Gamallo

Centro Singular de Investigación en Tecnoloxías da Información (CiTIUS),
Universidade de Santiago de Compostela, Santiago de Compostela, Spain
{marcos.garcia.gonzalez,pablo.gamallo}@usc.es

Abstract. This paper presents the current development of a multilingual suite for Natural Language Processing. It consists of a sentence chunker, a tokenizer, a PoS-tagger, a dictionary-based lemmatizer and a Named Entity Recognizer (both for *enamex* and *numex* expressions). The architecture of the pipeline and the main resources used for its development are described. Besides, the PoS-tagger and the Named Entity Recognizer are evaluated against several state-of-the-art systems. The experiments performed in Portuguese and English show that, in spite of its simplicity, our system competes with some well known tools for NLP. It is entirely written in Perl and distributed under a GPL license.

Keywords: Natural language processing · Pos-tagging · Named entity recognition · Portuguese · English

1 Introduction

This paper presents CitiusTools, a multilingual suite for Natural Language Processing (NLP) which performs the following tasks: sentence chunking, tokenization, PoS-tagging, lemmatization and Named Entity Recognition (NER). The suite is entirely written in Perl and distributed under a GPL license.[1]

The paper presents the architecture of the pipeline as well as its adaptation to Portuguese and English (the Spanish version was introduced in [6]). It is also presented a set of experiments aimed at knowing the performance of the PoS-tagger and the NE classifier modules. The results sho w that, in spite of its simplicity, our system behaves quite well when compared to some state-of-the-art suites such as Stanford CoreNLP or FreeLing. Besides, it performs notoriously better than the models provided by other systems such as OpenNLP.

Section 2 introduces some related work. Then, the architecture of the system is presented in Sect. 3. Section 4 shows the external resources used for its adaptation to Portuguese and English, while Sect. 5 contains the performed experiments. Finally, Sect. 6 describes the main conclusions of this paper.

---

[1] http://proxectos.citius.usc.es/hpcpln/index.php.

© Springer International Publishing Switzerland 2015
J.-L. Sierra-Rodríguez et al. (Eds.): SLATE 2015, CCIS 563, pp. 65–75, 2015.
DOI: 10.1007/978-3-319-27653-3_7

2   Related Work

In the last years, several open-source NLP suites have been published, being available to the users. Some of them provide models for languages such as Portuguese and English (evaluated in this paper), while others include analyzers for other varieties such as Spanish, Chinese, German or Arabic.

Stanford CoreNLP [11] is one of the best known suites, including modules like tokenizers, PoS-taggers, named entity recognizers, coreference resolution systems and syntactic parsers. It is written in Java and has been developed mainly for English, but recently there have been published models for languages such as Spanish, Chinese, German or Arabic.

FreeLing [12] is a suite of language analyzers (written in C++) which includes similar modules than the Stanford system, and also has tools for other tasks such as phonetic encoding. Most of FreeLing modules analyze data in Catalan, Spanish, Portuguese, English, or French (among others).

Another toolkit for NLP analysis written in Java is OpenNLP,[2] which performs most common NLP tasks. There are available models for several language for this system, including English, Spanish or German.

Finally, IXA pipes [1] (a modular set also written in Java) performs tokenization, PoS-tagging, NER and parsing. Among the languages covered by this tool (depending on the module) are Spanish, English, Basque, Italian or Galician.

The system presented in this paper is, to the best of our knowledge, the first one written entirely in Perl. It provides a simple, efficient and ready to use set of NLP tools with a performance close to the state-of-the-art.

3   Architecture

Our system consists of five modules that can be applied in a pipeline in order to perform NLP tasks. The current version contains the following tools:

3.1   Sentence Chunker

This module is composed of a language-dependent list of abbreviations and a set of Finite State Automata (FST) aimed at identifying sentence boundaries.

The automata detect entities such as urls, e-mail addresses, and other elements containing dots that are not in sentence-ending position. Also, abbreviations ending in a dot character (e.g., *Dr.*, *corp.*, etc.) are not marked as sentence boundaries (except if their context is covered by a FST).

The output of this module is the input text with one sentence per line.

3.2   Tokenizer

The next module of the suite splits each identified sentence into its tokens. It is a rule-based tokenizer enriched with few language-dependent adaptations.

---

[2] http://opennlp.apache.org/.

First, the tokenizer identifies compound punctuation (such as ellipsis) and other punctuation inside numerical expressions. After that, a simple blank-space tokenizer is applied (which also splits the punctuation which do not belong to larger expressions).

Then, a battery of language-dependent rules is applied in order to split contractions (e.g. *don't* > *do/not*, in English), verb+pronoun forms (e.g., *mantém-se* > *mantém/se*, in Portuguese) and other elements which are useful for further NLP analysis. Note that some forms can be ambiguous between a contracted and a non-contracted element: *desse*, in Portuguese, could be a single token form of the verb *dar* (to give), or a contracted form of a preposition and a demonstrative (*de/esse*). As the decision for splitting these forms depends on their PoS-tag, the tokenizer does not split them. Thus, as in other works [8], these forms are analyzed by the PoS-tagger, which will split them (or not), according to the selected PoS-tag. Those cases where an element of the contraction may represent two different tokens (with a different PoS-tag, e.g., *I'd* > *I would* or *I had*, in English) are also splitted in this step, but the lemma will be provided by the disambiguation of the PoS-tagger.

The output of this module is a vector of tokens representing each previously identified sentence.

3.3 PoS-Tagger

The PoS-tagger assigns a morphosyntactic tag (from a set of predefined tags, the *tagset*) to each token.

This module is a bayesian classifier based on bigrams of tokens. It uses additive smoothing, which is commonly a component of bayesian classifiers. In order to label a token, the classifier calculates the probability of each tag (t_i) linked to the token, taking into account a set of contextual features $A_i...A_n$:

$$P(t_i \mid A_1, ...A_n) = P(t_i) \prod_{i=0}^{N} P(A_i \mid t_i) \tag{1}$$

The best set of features, selected in preliminary tests, was the following:

- t_{i-1}: the PoS-tag of the previous token.
- t_{i+1}: the PoS-tag of the next token.
- (k_i, t_{i-1}): the cooccurrence of the ambiguous token k_i together with the tag of the previous token.
- (k_i, t_{i+1}): the coccurrecence of the ambiguous token k_i together with the tag of the next token.

The model needs to be trained with a labeled corpus and a dictionary with the possible PoS-tags for each known token. The algorithm disambiguates the tokens from left to right, so the left context of an ambiguous token is an already labeled one. Thus, the features concerning the tag of the next token (t_{i+1}) include the probabilities of the different tags that could be associated with this token.

This strategy is similar to the Hidden Markom Models (HMM) algorithm proposed in [2]. The main difference is that our system handles the PoS-tagging as an individual classification problem (token by token), instead of searching for the best sequence of PoS-tags. Its computational efficiency is the main reason for the use of this simple approach.

The tagsets of the PoS-taggers follow the EAGLES guidelines [10]. For Portuguese, it has been used a tagset with 193 elements. The English tagset has 27 tags. Both of them have 9 extra tags for punctuation. The difference between these tagsets come from the complex verbal conjugation and nominal inflection of Portuguese. However, note that the classifier does not use the 193 elements in Portuguese: it just uses 21 tags for disambiguating the morphosyntactic category (e.g., noun, adjective) of each word. The other information (gender, number, tense, etc.) is then taken from the labeled dictionary.

The output of the PoS-tagger is the input vector enriched with a morphosyntactic label for each token.

3.4   Named Entity Identifier

The next module of the pipeline is a FST identifier of *numex* and *enamex* (named entities) expressions.

Before starting the identification process, this module takes advantage of a lemmatized dictionary (see Sect. 4) in order to assign a lemma for each token. It also uses the predicted PoS-tag for disambiguating tokens with different lemmas depending on their morphosyntactic category.

For identifying *numex* expressions (in our system: dates, currencies, numbers, measures and quantities), it is applied a set of language-dependent FSTs that cover the most common forms of representing these elements in each language.

The named entities (*enamex* expressions) are identified taking into account both their capitalization and possible functional words inside them (e.g., *Banco de Portugal*). In order to better identify the boundaries of the *enamex* expressions, this module also needs a list of words which can be both common words at sentence beginning position and the first element of a named entity (e.g., *Neves*, which can be a capitalized noun and a proper noun —surname or location— in Portuguese). These ambiguous forms are obtained semi-automatically using dictionaries and lists of gazetteers.

The output of this module is the input vector enriched with the identification of the *numex* and *enamex* expressions, as well as with the lemmas provided by the dictionary.

3.5   Named Entity Classifier

The named entity classifier module assigns each *enamex* one of the following labels: *person, organization, location* or *misc* (miscellaneous).

In order to classify an entity, this module uses large lists of encyclopedic gazetteers together with a set of rules for semantic disambiguation.

The gazetteers were automatically extracted from structured resources such as Freebase[3] and DBpedia,[4] and enriched with semi-structured knowledge obtained from the infoboxes and category trees of Wikipedia.[5] The gazetteers consist in four lists of entities (one for each semantic category). Besides, the system also uses small lists of *trigger words*, which are nouns that can subclassify an entity (e.g., "singer" for the class *person* or "company" for *organization*). The trigger words were also automatically extracted from the category trees of Wikipedia. Finally, a list of the most frequent personal names for each language (which are not common nouns) is used.

Concerning the disambiguation rules, they are applied using the following strategy for each named entity:

1. If the entity appears only in one of the gazetteers lists, it is classified with the class it belongs to.
2. If the entity appears in several lists (or if it does not appear in any), the context is analyzed. This context includes two windows (*before* and *after*) of three tokens each. If a trigger word is found in the context, the entity is classified as belonging to the trigger word class (with some restrictions such as trigger words in preposition phrases. "Caixa Geral" will not be labeled as *person* even if the trigger word "president" occurs in the context: *president of Caixa Geral*).
3. If the entity starts (or is) a frequent personal name present in the list, it is classified as *person*.
4. If the entity is ambiguous (it appears in more than one list or contains trigger words from different classes) and it cannot be disambiguated by its context, it is selected the most probable class (*prior probability*), by computing the distribution of the gazetteers in the Wikipedia.
5. If the context is not enough to disambiguate the entity, a rule verifies whether it contains a trigger word or the first token of a gazetteer inside. If there are more than one option, the gazetteers are preferred over the trigger words, and in case of ambiguity the prior probability is also computed.
6. If the previous rules cannot classify the entity, it is labeled as *misc*.

Note that the rules are mainly language-independent. In our case, only one rule had to be changed when adapting the system for English: a trigger word inside an entity appears in final position, instead of in the beginning, as in Portuguese (*National **Museum** versus **Museu** Nacional*).

Even though the performance of this module depends on the quality and persistence of the gazetteers, the use of contextual features together with the combination of rules that analyze the internal form of each entity allow the system to keep reasonable accuracy even in unknown forms.

---

[3] http://www.freebase.com.

[4] http://www.dbpedia.org.

[5] http://www.wikipedia.org.

Table 1. Summary of the size of the resources: dictionaries, PoS-tagger training corpora, NER testing corpora and total number of gazetters.

Language	Dictionary	PoS-tagger (train)	NER (test)	Gazetteers
Portuguese	1.250 M	130 k	75 k	100 k
English	350 k	1 M	524 k	1.5 M

4   Resources

This section briefly describes the external resources used by the different NLP modules of our system. Table 1 includes a summary of these data.

4.1   Portuguese

For training the PoS-tagger for Portuguese (and also for extracting some lists described above), we used the dictionary of FreeLing based on the Label-Lex lexicon [4]. It consists of ≈ 1.250 million pairs token-tag from about 120 k lemmas.

For training the PoS-tagging we used a subset of the CoNLL version of the Bosque 8.0, with about 130 k tokens.[6] For testing, we used a different subset of the Bosque and three small corpora of European Portuguese (EP) news, Brazilian Portuguese (BP) news and a Wikipedia articles.

For testing the named entity classification, there were used both a subset of the labeled version of the Bosque (≈ 20 k tokens) and the Corpus-Web (with about 55 k tokens of different varieties of Portuguese) [9].

In order to build the gazetteers, the Portuguese version of the Wikipedia was used for extracting entity names. Apart from that, large lists of countries and cities were also merged, together with the most common names and surnames in Portuguese an other lists of gazetteers freely available (such as the FreeLing data), generating the following lists: 59, 421 *person* entities, 14, 197 *organizations*, 34, 590 *locations* and 838 for *misc* gazetteers.

4.2   English

For English, the morph_english dict.v1.4 was used, with about 350 k token-tag pairs from ≈ 77.5 k lemmas.[7] For training and testing the PoS-tagger we used the Brown corpus, with ≈ 1.2 million tokens:[8] ≈ 1 million tokens were randomly selected for training, while the tests were carried out with the other 200 k tokens. Both the dictionary and the corpora had to be adapted and converted to the same tagset.

The classification of named entities was evaluated using two corpora: the IEER,[9] with 68, 402 tokens and classification of *person*, *location* and *organization*

---

[6] http://www.linguateca.pt/floresta/CoNLL-X/.

[7] ftp://ftp.cis.upenn.edu/pub/xtag/morph-1.5/morph-1.5.tar.gz.

[8] http://clu.uni.no/icame/brown/bcm.html.

[9] http://www.itl.nist.gov/iad/894.01/tests/ie-er/er_99/er_99.htm.

entities (not *misc*), and the SemCor Corpus,[10] with a size of 455, 597 tokens and annotation of the four *enamex* classes. The PoS-tags of this last corpus had been predicted (not manually revised).

The English gazetteers were extracted from Freebase and DBpedia, enriched with lists of countries and capitals and the most common names and surnames in this language. The final versions had the following size: 922, 767 for *person*, 126, 334 for *organization*, 351, 151 for *location* and 94, 525 for *misc*.

5   Evaluation

This section describes the evaluation experiments performed on the two main modules of the system: the PoS-tagger (CitiusTagger) and the NE classifier (CitiusNEC). The experiments were carried out in Portuguese and English, using three NLP suites for comparison: FreeLing (for Portuguese), and Apache OpenNLP and Stanford CoreNLP (for English).[11]

It is important to note that some results are not strictly comparable, since we used the models provided by each software. On the one hand, these models were trained with different resources (corpora, lexicons, gazetteers...), having also different tagsets (quickly adapted for doing the experiments). On the other hand, the alignment between the gold-standard and the test files also involved variation on the results (as it is shown below).

So, the objective of this evaluation is not to know what is the best system for PoS-tagging and NE classifying texts in Portuguese and English, but to have a decent comparison of our system analyzing the same data as other NLP suites.

5.1   PoS-Tagger

The first set of experiments compared the performance of the PoS-tagger in Portuguese and English.

Table 2. PoS-tagging results (precision) for Portuguese.

Corpus	Size	CitiusTagger	FreeLing
Bosque	80,881	96.07	96.62
EP News	13,964	96.70	97.76
BP News	11,476	95.73	96.99
Wikipedia	17,149	95.76	96.13
Macro-average	—	96.06	96.88
Micro-average	—	96.06	96.72

---

[10] http://www.gabormelli.com/RKB/SemCor_Corpus.

[11] The output of each system as well as the gold-standard files can be obtained in the following url: http://gramatica.usc.es/~marcos/slate15.zip.

Table 2 contains the results for Portuguese. Our bayesian PoS-tagger were compared to the HMM model of FreeLing [8,12], analyzing the four mentioned corpora (see Sect. 4). The results include the precision (true positives/true positives + false negatives) on each corpora as well as the macro and micro-average values (macro-average is the harmonic mean of the results from each corpus while micro-average values are computed from the sum of all the true and false positives and negatives from each corpora).

When compared to the HMM model, our system behaves quite similar in every corpora (with a maximum difference of −1.2 in BP News), with average results of 96 % precision. Note that this comparison is strict, since both the gold-standard and the testing corpora were perfectly aligned. Besides, the tagset of our system and the FreeLing one were almost identical.

In English, the bayesian PoS-tagger was compared to three different models (in one corpus): the maximum entropy and perceptron classifiers of OpenNLP (1 and 2, respectively) and the Stanford POS Tagger (maximum entropy) [13].

The output of the external systems (OpenNLP and Stanford) were automatically converted to the same tagset of the gold-standard.

Table 3. PoS-tagging results (precision) for English. OpenNLP_1 is a maximum entropy model, while OpenNLP_2 is a perceptron classifier. Test corpus has a size of 209, 406 tokens.

CitiusTagger	OpenNLP_1	OpenNLP_2	Stanford
93.55	91.72	90.93	91.12

The results (Table 3) show that our PoS-tagger behaves as good as the maximum entropy and perceptron models. Actually, the precision of the bayesian model is almost 2 % higher, but the evaluation cannot be strict: some minority tags (e.g. FW for *foreign words*) appeared in the gold-standard but not in the tagsets of these taggers (and vice versa).

However, these experiments (together with the Portuguese ones) suggest that the bayesian model achieves a high performance despite its simplicity.

5.2   Named Entity Classifier

Concerning NE classification, the Portuguese system was also compared to the FreeLing AdaBoost classifier [3,7] in two corpora: Bosque and Corpus-Web.

Table 4 shows the results of these two classifiers in the referred corpora. In Bosque, our system achieved slightly better results than the AdaBoost classifier, while in Corpus-Web, the FreeLing module had better results.

Again, the average results show that a simple system (based on resources and rules) has similar performance than a supervised classifier.

Table 4. Named entity classification results (f-score) for Portuguese. *NEs* refers to the number of full *enamex* entities (not tokens) in each corpus.

Corpus	Tokens	NEs	CitiusNEC	FreeLing
Bosque	19,579	1,027	90.07	88,89
Corpus-Web	55,305	3,666	73.76	75.31
Micro-average	—	—	81.92	82.10
Macro-average	—	—	77.33	78.22

Table 5. Named entity classification results (f-score) for English.

Corpus	Tokens	NEs	CitiusNEC	OpenNLP	Stanford
IEER	68,402	3,384	75.95	52.77	75.86
SemCor	455,597	9,696	58.81	44.85	65.57
Macro-average	—	—	63.38	48.90	70.72
Micro-average	—	—	63.23	47.10	68.63

In English, the resource-based method was compared to the OpenNLP (Name Finder models)[12] and to the Stanford NER (CRF with distributional similarity features in an IOB2 classifier)[13] [5].

The output of these systems were automatically converted to the CoNLL IOB format (used in both versions of the IEER and SemCor corpora).

The results of the named entity classifiers (Table 5) show that in the IEER corpus, our system behaves as good as the Stanford model, while in SemCor, the former increased our performance in more than 7 %. In average, our resource-based classifier had much better performance ($\approx$ 5 %) than the OpenNLP system, while the Stanford one increased our results in 5 % − 7 % f-score.

Finally, it was carried out a test aimed at knowing the processing speed of the evaluated systems. They were used for labelling a Spanish corpus of 100,000 tokens (in an Intel Core 2 2.5 GHz processor with 4 gb of RAM running Debian Jessie). The systems needed the following time for applying the pipeline (sentence chunker, tokenizer, PoS-tagger and NER): OpenNLP (only NER): 1 m 48 s; FreeLing: 2 m 27 s; CitiusTools: 2 m 38 s and Stanford CoreNLP: 11 m 25 s.

In sum, the evaluations performed with the two main modules of our pipeline—CitiusTagger and CitiusNEC—suggest that they achieve very good results (some of them comparable to state-of-the-art systems) despite their simplicity and their quick adaptation to Portuguese and English. This is in accordance with the results obtained for Spanish, such as it was described in [6].

---

[12] http://opennlp.sourceforge.net/models/english/namefind/.
[13] http://nlp.stanford.edu/software/conll.distsim.iob2.crf.ser.gz.

6   Conclusions and Further Work

This paper presented the current version of CitiusTools, a multilingual suite for NLP which includes modules for the most common tasks of this field.

The modules, written in Perl, combine some rule-based and supervised models which take advantage of external resources such as lexicons, labeled corpora or large lists of gazetteers.

Two different modules (PoS-tagger and NER) were evaluated in Portuguese and English, compared to some of the best NLP tools available for these languages. The results showed that the performance of our system is similar than the state-of-the-art, even if it has been quickly adapted to these languages.

In current work, we are adapting all the modules in the suite to two new languages (Galician and French), and we expect to include (in further work) a deterministic module for coreference resolution.

References

1. Agerri, R., Bermudez, J., Rigau, G.: IXA pipeline: efficient and ready to use multilingual NLP tools. In: Proceedings of the 9th Language Resources and Evaluation Conference (LREC 2014), Reykjavik (2014)
2. Brants, T.: TnT - a statistical part-of-speech tagger. In: Proceedings of the 6th Conference on Applied Natural Language Processing (ANLP). Association for Computational Linguistics (2000)
3. Carreras, X., Màrquez, Ll., Padró, Ll.: A simple named entity extractor using adaboost. In: Proceedings of the Conference on Natural Language Learning (CoNLL 2003) Shared Task. Edmonton (2003)
4. Eleutério, S., Ranchhod, E., Mota, C., Carvalho, P.: Dicionários electrónicos do Português. Características e Aplicações. In: Actas del VIII Simposio Internacional de Comunicación Social, pp. 636–642, Santiago de Cuba (2003)
5. Finkel, J., Grenager, T., Manning, C.: Incorporating non-local information into information extraction systems by gibbs sampling. In: Proceedings of the 43rd Annual Meeting of the Association for Computational Linguistics (ACL 2005), pp. 363–370 (2005)
6. Gamallo, P., Pichel, J.C., Garcia, M., Abuín, J.M., Pena, T.F.: Análisis morfosintáctico y clasificación de entidades nombradas en un entorno big data. Procesamiento Lenguaje Nat. **53**, 17–24 (2014)
7. Garcia, M.: Extracção de Relações Semânticas. Recursos, Ferramentas e Estratégias. Ph.D. thesis, University of Santiago de Compostela (2014)
8. Garcia, M., Gamallo, P.: Análise Morfossintáctica para Português Europeu e Galego: Problemas, Soluções e Avaliação. LinguaMÁTICA **2**(2), 59–67 (2010)
9. Garcia, M., Gamallo, P.: Multilingual corpora with coreferential annotation of person entities. In: Proceedings of the 9th edition of the Language Resources and Evaluation Conference (LREC 2014), pp. 3229–3233, Reykjavik (2014)
10. Leach, G., Wilson, A.: Recommendations for the morphosyntactic annotation of corpora. Expert Advisory Group on Language Engineering Standard, Techincal report, EAGLES (1996)

11. Manning, C., Surdeanu, M., Bauer, J., Finkel, J., Bethard, S.J., McClosky, D.: The stanford CoreNLP natural language processing toolkit. In: Proceedings of 52nd Annual Meeting of the Association for Computational Linguistics (ACL 2014): System Demonstrations, pp. 55–60 (2014)
12. Padró, L.I.: Analizadores multilingües en freeling. LinguaMÁTICA **3**(2), 13–20 (2011)
13. Toutanova, K., Klein, D., Manning, C., Singer, Y.: Feature-Rich Part-of-Speech tagging with a cyclic dependency network. In: Proceedings of the Human Language Technology and the North American Chapter of the Association for Computational Linguistics (HLT-NAACL 2003), pp. 252–259, Edmonton (2003)

Human-Computer Languages

Towards a DSL for Educational Data Mining

Alfonso de la Vega, Diego García-Saiz, Marta Zorrilla, and Pablo Sánchez(⊠)

Dpto. Ingeniería Informática y Electrónica,
Universidad de Cantabria, Santander, Spain
{alfonso.delavega,diego.garcia,marta.zorrilla,p.sanchez}@unican.es

Abstract. Nowadays, most companies and organizations rely on computer systems to run their work processes. Therefore, the analysis of how these systems are used can be an important source of information to improve these work processes. In the era of Big Data, this is perfectly feasible with current state-of-art data analysis tools. Nevertheless, these data analysis tools cannot be used by general users, as they require a deep and sound knowledge of the algorithms and techniques they implement. In other areas of computer science, domain-specific languages have been created to abstract users from low level details of complex technologies. Therefore, we believe the same solution could be applied for data analysis tools. This article explores this hypothesis by creating a Domain-Specific Language (DSL) for the educational domain.

Keywords: Domain-specific languages · Big data · Educational data mining

1 Introduction

Nowadays, most work processes in companies and organizations are supported by software systems. Thus, the way in which people interact with these systems reflects somehow how these processes are actually executed. Therefore, a careful analysis of this interaction can help to find out flaws of these processes that might be removed [4].

For instance, let us suppose a company which wants to reduce the number of products that are returned after having been shipped. In this scenario, the company managers might be interested in getting answers for questions such as: *"What features share those products that are returned by customers?"*; or *"What is the profile of the unsatisfied customers?"*. Decision makers need to know these answers before adopting corrective actions.

Currently, it is feasible to perform this data analysis by using Big Data technologies [3]. For instance, the profiles of the unsatisfied customers can be computed using clustering techniques [8]. These techniques require a sound knowledge of the algorithms and mathematical foundations they use. Nevertheless, average decisions makers do not have this knowledge.

For example, to execute a clustering, the user should know how a clustering algorithm like *K-means* [1] works, how its parameters must be configured,

© Springer International Publishing Switzerland 2015
J.-L. Sierra-Rodríguez et al. (Eds.): SLATE 2015, CCIS 563, pp. 79–90, 2015.
DOI: 10.1007/978-3-319-27653-3_8

or what are the advantages of *K-means* as compared to other clustering algorithms, such as *X-means* [13]. Since decision makers lack of this knowledge, they need to rely on third-parties to carry out these data analysis processes, which leads to a costs increment and a productivity reduction.

In other areas of software development, *Domain Specific Languages (DSLs)* [9,18] have been created in order to allow users without expertise in a certain technology to use it. This is achieved by abstracting low level details of the underlying technology and by using a syntax and a terminology familiar to the end-user.

Therefore, we propose to build DSLs for data analysis. These DSLs would allow decision makers to formulate queries about the performance of a business process using a syntax and terminology familiar to them. Then, these queries would be automatically transformed into invocations of specific algorithms for data analysis. The DSL syntax should hide all the details associated with data analysis techniques to the end-user, who might remain unaware of how these techniques are used.

This article explores the feasibility of this idea by showing how a DSL with these characteristics can be developed for the e-learning domain. The objective of this DSL is to analyze the performance of a course hosted on an e-learning platform, such as *Moodle*, by using data, like the students' activity, gathered via this kind of platform. The final users of the DSL will be teachers and instructors, so it must use a syntax and a terminology familiar to them. Similar DSLs might be created for other domains following the process described in this article.

For the development of the DSL, we will make use of modern model-driven engineering techniques. More specifically, we will follow the development process proposed by Kleppe [9].

After this introduction, this article is structured as follows: Sect. 2 describes the domain our DSL targets. Section 3 comments on related work. Sections 4 and 5 explain how a DSL for the educational domain has been developed. Finally, Sect. 6 discusses the benefits of this work and concludes this article.

2   Educational Data Mining

The first step to develop a DSL, according to [12], is to know for what purpose the DSL will be used and obtain a sound knowledge of the domain it will target. In our case, this domain will be the educational domain, and we are mainly interested in knowing for what kind of questions decision makers would like to get an answer. Moreover, we are also interested in discovering what data is available to compute these answers. The domain information has been obtained using our own experience in the educational data mining domain as well as the assistance of several external teachers and instructors. In the following, the Educational Data Mining domain is described briefly.

Data Mining is the process of discovering interesting patterns and knowledge from large amounts of data [8]. In the last few years, it has been applied to the educational domain, what is known as *Educational Data Mining (EDM)* [17]. Educational Data Mining aims to take advantage of the data gathered by e-learning platforms, such as *BlackBoard* [16] or *Moodle* [15], which store data related to the activity carried out by the students of their courses. Educational Data Mining is defined as *"an emerging discipline, concerned with developing methods for exploring the unique types of data that come from educational settings, and using those methods to better understand students, and the settings which they learn in."* [17]. The discovered information could be useful for teachers and instructors in order to improve the performance of their teaching-learning processes.

For instance, at the beginning of a course, a teacher might be interested in what kind of students' profiles exist. Based on the obtained information, the teacher might adapt the course before it starts in order to tune it for these students. Thus, at the beginning of the course, the teacher could ask: *"What are the profiles of my students?"*. This information can be computed by using *clustering techniques* [8] on the students' demographic and activity data.

When the course finishes, teachers are usually worried about the students that have not passed the course. Therefore, they would like to refine the previous question and ask: *"What are the profiles of the students who have not passed?"*. As before, this information can be computed using clustering techniques on the students' data, but removing those students that have passed from this dataset. Moreover, teachers are obviously interested in asking *"What are the reasons why my students failed?"*. This might be partially answered by applying *classification techniques* [8] on the students' data, by analysing the student activity logs to find out these reasons.

Obviously, most teachers know nothing about clustering and classification rules, so they cannot use these techniques directly by themselves. This is the reason behind the aim of hiding these details to the end-user.

Next section analyses whether this objective can be achieved using current state-of-art techniques.

3   Related Work

To the best of our knowledge, there is little work done about how to make data analysis techniques more usable by decision makers. The approaches which tackle this issue can be grouped in two sets.

The first group aims to assist decision makers in the process of defining a data analysis process. For instance, [5] defines a method where end-users are prompted with different questions, which guide them in the definition of a data mining process that fits in with their needs.

For example, a question could be if the decision maker is interested in computing the profiles of a certain dataset. If so, the user is asked for more detailed information that is required to execute this task. Some of these questions might result confusing. As an example, the user should be able to answer about how a certain data is represented, if as a string label or as a numerical value. An average decision maker might not know these technical details. Moreover, answering these questions can be a large and tedious process, which could lead to build wrong mining models or to stop using the tool.

In [2], a query-by-example based language is defined. In basic query-by-example, the decision maker constructs a prototype of an answer for the question he or she would like to ask. This prototype is a table, where each column represents an attribute of the desired answer. These columns can be constrained to certain values, which are used to select the desired results. The system depicted in [2] enhances this table with specific columns to execute data mining processes. Again, the information we need to supply in these columns requires some knowledge of the underlying data analysis technique to be applied, so the user is not completely unaware of these techniques. Moreover, the construction of these prototypes is based on data warehouse concepts, such as OLAP (*On-Line Analytical Processing*) [20]. Average decision makers also often lack of this kind of knowledge.

In the second group, there are software applications with prebuilt data mining processes which can be directly executed by decision makers. An example of this strategy is *E-learning Web-Miner (ElWM)* [21]. ElWM is a web-based application whose objective is to allow instructors to analyse the performance of a course hosted in an e-learning platform. At the time of writing this article, ElWM offers instructors answers to three different queries: (1) what kinds of resources are frequently used together (e.g., forum, mail) in each learning session; (2) what are the profiles of the different sessions carried out by students; and (3) what are the profiles of the students enrolled in a course.

In this case, the main limitation is that the set of queries is fixed and they cannot be refined without modifying the application. For instance, if we wanted to compute the profiles of assignments which students have failed; or the profiles of students that do not pass the course, we would need to update the application to allow this more specific filtering.

By developing DSLs for data analysis, we expect to overcome these shortcomings. Next sections describe how this task is accomplished for the educational domain.

4   Grammar Specification

As previously commented, we will follow the process proposed by Kleppe [9] for the development of the DSL. According to this process, the first step to implement a DSL, once the knowledge about the target domain has been collected, is to specify its grammar. Next subsections describe how this step is accomplished.

The definition of a grammar for a DSL, following a model-driven perspective, implies the definition of an abstract syntax and a concrete syntax. The abstract syntax specifies the grammar of a language independently of how this model is represented. The concrete syntax is a specific rendering, either textual or visual, for the abstract syntax. We describe both elements below.

4.1   Abstract Syntax

Abstract syntaxes are usually specified by using *metamodels* [11]. A metamodel can be considered as a model of the syntax of a language. For the construction of

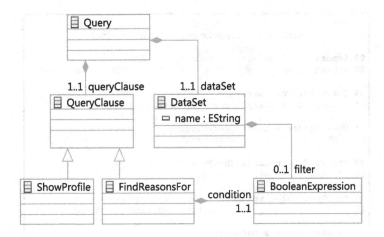

Fig. 1. Abstract syntax of our DSL

this metamodel, we have used Ecore [19], which is the de-facto standard language for metamodeling. Figure 1 shows the metamodel for our DSL syntax.

According to this metamodel, our language allows us to write queries. A *Query* has a *QueryClause*. In the figure, two query clauses are depicted: *ShowProfile* and *FindReasonsFor*. A query clause can be viewed as a command that hides a data mining technique. Moreover, each query has an associated *DataSet*, which must be available in a well-defined location.

Moreover, a data source can have an associated *filter*. A *filter* is a boolean expression that selects the subset of instances of a data source which satisfies such expression. The abstract syntax for boolean expressions are not shown in this article for the sake of simplicity and brevity, as this syntax is probably known by the reader.

Filters are used to apply a query clause to a specific subset of a data source. For instance, an instructor might be interested in selecting students that: (1) do not pass a course; (2) drop out; or (3) are above or below a certain age, among other options. Obviously, these filters must be written using the attributes of the data source. For instance, if students' age is not stored in the database, it could not be used in a filter.

In the case of the *FindReasonsFor* clause, an additional *condition* is required because the goal of this query is to compute the reason why certain instances of the data source satisfy a certain condition. As before, this condition is a boolean expression.

After developing the abstract syntax of our DSL, the next step is to specify its concrete syntax, which is described in the following subsection.

4.2 Concrete Syntax

We have opted for developing a textual syntax for our DSL. However, this issue needs to be further investigated, as some decision makers might prefer a graphical notation.

```
00 grammar es.unican.dslEdm.Dsl
01         with org.eclipse.xtext.common.Terminals

02 import "EdDataMiningMetamodel"
03 import "http://www.eclipse.org/emf/2002/Ecore" as ecore

04 Query returns Query:
05   queryClause=QueryClause 'of' dataSet=DataSet;

06 QueryClause returns QueryClause:
07   ShowProfile | FindReasonsFor;

08 ShowProfile returns ShowProfile:
09   'show_profile';

10 FindReasonsFor returns FindReasonsFor:
11   'find_reasons_for' condition=BooleanExpression;

12 DataSet returns DataSet:
13   name=ID ('with' filter=BooleanExpression)?;

...
```

Fig. 2. Textual concrete syntax for our DSL

For the definition of the textual concrete syntax, we have used *Xtext* [6], which allows DSL developers to define a textual syntax from a Ecore metamodel. Using *Xtext*, a grammar is defined following a notation similar to EBNF (*Extended Backus–Naur Form*), but where the production rules are enhanced with constructions to create instances of metaclasses from the metamodel as the grammar is parsed.

Figure 2 shows the concrete syntax for our DSL. Lines 00–03 specify: (1) the namespace and name for the grammar; (2) include a convenience package called *Terminals*, provided by *Xtext*; and, (3) specifies that the grammar will be based on the Ecore and the *EdDataMiningMetamodel* metamodel, which corresponds to the abstract syntax depicted in Fig. 1.

Then, Lines 04–05 specify *Query* as the entry point of our grammar. A *Query* is composed of a *QueryClause*, followed by the *of* keyword and the specification of a *Dataset*. Both lines are equivalent to the EBNF rule *Query* :: = *QueryClause* *"of"* *DataSet*.

Moreover, in Line 04, the *return Query* clause specifies that an instance of the *Query* metaclass (see Fig. 1) must be created when this production rule is executed. Furthermore, the results of executing the *DataSet* and *QueryClause* production rules must be assigned to the *queryClause* and *dataSet* attributes of the *Query* metaclass, respectively.

Similarly, a *QueryClause* can be either a *ShowProfile* or a *FindReasonsFor* (Lines 06–07) clause. In the first case, the query clause is simply written using the *show_profile* keyword (Lines 08–09). In the second case, after the keyword *find_reasons_for*, a boolean expression that serves as *condition* for evaluating the query is required (Lines 10–11).

```
00 show_profile of Students;

01 show_profile of Students with courseOutcome=fail;

02 find_reasons_for courseOutcome=fail of Students;
```

Fig. 3. Queries written using the DSL

Finally, a *Dataset* is simply represented by an identifier plus an optional filter definition (Lines 12–13). This identifier must correspond to an available dataset. This constraint is checked by means of external rules.

Once the grammar has been specified, a full editor for our grammar, with syntax colouring, helpers and automatic formatting, as well as parsing, type-checking and validation capabilities can be automatically generated by *Xtext*. Using this editor, queries as shown in Fig. 3 can be written.

Thus, instructors can now write queries to analyse course performance by using a terminology that is familiar to them. The next step is to provide execution capabilities to these queries, which is achieved by translating them into Java code.

5   Query Execution

To compute the result of a query, data mining techniques are used. For instance, to identify profiles in a dataset, clustering techniques must be chosen. Therefore, the strategy to execute a query is to transform it into a Java code snippet which invokes a prebuilt implementation of the corresponding data mining algorithm. In our case, these prebuilt implementations are provided by *Weka* [7], a widely used data mining tool suite.

It should be taken into account that most data mining algorithms require the specification of a set of input parameters, which are necessary for tuning the algorithm. For instance, most clustering techniques require the specification of the number of clusters to be built. Obviously, if the ultimate goal of the DSL is to abstract the end user from data mining techniques, it cannot be expected that the end-user provides the values for these parameters.

Therefore, these parameters have to be self-computed. Currently, there is a research area inside the data mining field, known as *parameter-less data mining*, that aims to build self-configuring data-mining algorithms. Thus, these techniques will be used whenever possible.

To illustrate how this code generation process works, we describe how the *Show Profile* queries are transformed into Java code. The query *show_profile of Students with courseOutcome=fail;* will be used as an example.

The code generation process has been implemented using templates. More specifically, we have used EGL (*Epsilon Generation Language*) [10], which is a language that allows code generation from Ecore-based models.

```
00  DataSource source = new DataSource("[%=query.dataSet.name%].arff");
01  Instances ins = source.getDataSet();
02  Instances insFiltered = ins;
    [%
03  if (query.dataSet.filter.isDefined()) {
04    var filterConstraints = query.dataSet.filter.operations;
05    for (constraint in filterConstraints) { %]
06          [%=constraint.toFilter()%]
07  [%}
08  }
    %]
09  XMeans xm = new XMeans();
10  xm.setMinNumClusters(Clustering.XMEANS_MIN_NUM_CLUSTERS);
11  xm.setMaxNumClusters(Clustering.XMEANS_MAX_NUM_CLUSTERS);
12  xm.buildClusterer(insFiltered);
13  Visualizer.saveClusterer(xm, Visualizer.ClusterType.XMEANS);
```

Fig. 4. General clustering template

```
00  [%@template
01    operation EqualityComparison toFilter() {
02      var attrName = "attr" + self.attrName;
03      var valueName = "value" + self.attrName; %]
04      Attribute [%=attrName%] = ins.attribute("[%=self.attrName%]");
05      String [%=valueName%] = "[%=self.value%]";
06      [% var filterName = "filter" + self.attrName; %]
07      RemoveWithValues [%=filterName%] = new RemoveWithValues();
08      [%=filterName%].setAttributeIndex(
09          Integer.toString([%=attrName%].index() + 1));
10      [%=filterName%].setNominalIndices(
11          Integer.toString([%=attrName%].indexOfValue([%=valueName%]) + 1));
12      [%=filterName%].setInputFormat(ins);
13      [%=filterName%].setInvertSelection(true); // matching entries
14      insFiltered = Filter.useFilter(insFiltered, [%=filterName%]);
15  [%} %]
```

Fig. 5. Filter for equality comparisons

As a first step, a template is selected based on the query clause. A fragment of the template applied for the *Show Profile* queries can be seen in Fig. 4, whereas the code generated by this template for our example query is shown in Fig. 6. The code generation process based on this template works as follows:

1. First, the dataset to be analysed is loaded, using the corresponding Weka helper classes. In this case, Lines 00–01 of Fig. 4 are in charge of generating Lines 04–05 of Fig. 6. The name of the dataset is obtained from the attribute name of the *Dataset* metaclass (Fig. 4, Line 00). As previously indicated, the parser must have checked that a dataset with that name exists.

2. In case the dataset has an associated filter, the code to perform this filtering must be generated. Consequently, the boolean expression which defines the filter must be transformed into the corresponding Weka code to filter a dataset according to the values of certain attributes. Figure 4, Lines 03–08

```
00 package processes;
01 import weka.[...]

02 public class ClusteringSnippet {
03   public static void main(String[] args) throws Exception {

04     DataSource source = new DataSource("Students.arff");
05     Instances ins = source.getDataSet();

06     Instances insFiltered = ins;
07     Attribute attrcourseOutcome = ins.attribute("courseOutcome");
08     String valuecourseOutcome = "fail";
09     RemoveWithValues filtercourseOutcome = new RemoveWithValues();
10     filtercourseOutcome.setAttributeIndex(Integer
                .toString(attrcourseOutcome.index() + 1));
11     filtercourseOutcome
                .setNominalIndices(Integer.toString(attrcourseOutcome
                        .indexOfValue(valuecourseOutcome) + 1));
12     filtercourseOutcome.setInputFormat(ins);
13     filtercourseOutcome.setInvertSelection(true); // matching entries
14     insFiltered = Filter.useFilter(insFiltered, filtercourseOutcome);

15     XMeans xm = new XMeans();
16     xm.setMinNumClusters(Clustering.XMEANS_MIN_NUM_CLUSTERS);
17     xm.setMaxNumClusters(Clustering.XMEANS_MAX_NUM_CLUSTERS);
18     xm.buildClusterer(insFiltered);
19     Visualizer.saveClusterer(xm, Visualizer.ClusterType.XMEANS);
   }
 }
```

Fig. 6. Code generated after processing the example query

show the template code which processes the constraints of the boolean expression and converts them to Java code. For each constraint, the *toFilter* method is invoked. The implementation of this method is different for each kind of constraint. In our example, an equality comparison is used to filter those students whose *courseOutcome* is *fail*. Therefore, the *toFilter* method corresponding to equality comparisons is invoked. The code generation template corresponding to this operation is shown in Fig. 5. This template makes use of the existent Weka filter *RemoveWithValues* to achieve this goal. Figure 6, Lines 07–14 show the resulting code for our concrete example.

3. Then, the code for executing the clustering algorithm on the loaded dataset is generated (Fig. 4, Lines 09–12, Fig. 6, Lines 15–18). In our case, the *X-means* algorithm [13] is selected to perform the clustering. The advantage of this algorithm is that it can estimate the number of clusters that should be created for a particular dataset. *X-means* only requires that this number is bound to a certain range. Therefore, if the lower and upper bounds of this range are set to proper values, the *X-means* algorithm can be used as a self-configuring algorithm. Since teachers expect to find at least two different students groups, 2 is a reasonable lower bound in this case. For a normal course, 20 is a number of clusters high enough to be considered as infinite, thus it is a reasonable upper bound for our algorithm. Therefore, the responsibility of determining

the number of clusters to be created is delegated into the *X-means* algorithm, which automatically calculates it.

4. Finally, Fig. 6, Line 13 shows how the result of the *X-means* algorithm is placed in an output file, which is read by a visualization tool in order to adequately render the results in a user-friendly way.

The code generation process for the *FindReasonsFor* queries would be similar, but in this case, a classification algorithm would be invoked, precisely, the *J48* implementation provided by Weka of the *C4.5* decision tree algorithm [14]. Therefore, a different EGL template would be used in this case.

With this last step, the development of our DSL for Educational Data Mining is finished. DSLs for applying data mining techniques in other domains might be developed following a similar process. Several excerpts of the DSL are agnostic of the target domain, thus they could be reused and, consequently, a remarkable reduction in the cost and development effort of a new DSL could be achieved.

Next section discusses whether this DSL satisfies the objectives of this work and concludes this article.

6   Conclusions

This article has shown how a Domain-Specific Language for Educational Data Mining can be developed. This DSL allows teacher and instructors of courses hosted in e-learning platforms to analyse the performance of their teaching-learning processes by means of applying data mining techniques on the data contained in such a platform. The DSL approach provides two benefits as compared to current state-of-art techniques.

First, the DSL abstracts low level-details of data analysis techniques, so it can be used by instructors without any knowledge of these techniques. Thus, our approach offers a solution to bridge the gap between data analysis tools and decision makers. The DSL syntax only contains high-level keywords and references to entities and attributes of the target domain data model. Thus, the DSL contains a terminology that should be known by the decision makers, who would be instructors and teachers in the case of the educational domain.

Secondly, the DSL is flexible enough to support the elaboration of arbitrary complex new queries. This is an advantage as compared to approaches that develop tools able to compute concrete tasks. ElWM [21] is an example of such a tool. As commented in Sect. 3, using ElWM students' profiles of a course can be computed. However, the profiles of students above a certain age cannot be computed without modifying the application. Similarly, we cannot calculate profiles of other entities, such as assignments, without updating the tool. This is, each time we want to modify a query, the application must be updated to support it.

Oppositely, the DSL offers a more flexible interaction. There exists limitations, as decision makers cannot ask any arbitrary question and they must adhere to the available query clauses. Thus, the included set of clauses should cover the potential questions decision makers are interested in asking. Moreover, the queries must be written following the syntactic rules of the grammar, as the necessity to parse them with a computer prevents the usage of most informal natural language expressions.

As future work, we expect to add more query options to the DSL for the educational domain, as well as to develop DSLs for other domains. More specifically, we are interested in developing DSLs for the performance analysis of work processes in the public administration.

Acknowledgements. This work has been partially funded by the Government of Cantabria (Spain) under the doctoral studentship program from the University of Cantabria, and the Spanish Government and FEDER funder under grant TIN2011-28567-C03-02 (HI-PARTES).

References

1. Arthur, D., Vassilvitskii, S.: K-means++: the advantages of careful seeding. In: Proceedings of the 18th Annual ACM-SIAM Symposium on Discrete Algorithms (SODA), pp. 1027–1035, New Orleans (Louisiana, USA), January 2007
2. Azevedo, A., Santos, M.: Binding data mining to final business users of business intelligence systems. In: 1st International Conference on Intelligent Systems 'and Applications (Intelli), pp. 7–12, April–May 2012
3. Baesens, B.: Analytics in a Big Data World: The Essential Guide to Data Science and Its Application. Wiley, New York (2014)
4. Bughin, J., Chui, M., Manyika, J.: Clouds, big data and smart assets: ten tech-enabled business trendsto watch. McKinsey Q. **56**, 1–14 (2010)
5. Espinosa, R., García-Saiz, D., Zorrilla, M., Zubcoff, J.J., Mazón, J.-N.: Enabling non-expert users to apply data mining for bridging the big data divide. In: Ceravolo, P., Accorsi, R., Cudre-Mauroux, P. (eds.) SIMPDA 2013. LNBIP, vol. 203, pp. 65–86. Springer, Heidelberg (2015)
6. Eysholdt, M., Behrens, H.: Xtext: implement your language faster than the quick and dirty way. In: Companion to the 25th Annual Conference on Object-Oriented Programming, Systems, Languages, and Applications (SPLASH/OOPSLA), pp. 307–309, Reno/Tahoe (Nevada, USA), October 2010
7. Hall, M., Frank, E., Holmes, G., Pfahringer, B., Reutemann, P., Witten, I.H.: The WEKA data mining software: an update. SIGKDD Explor. Newsl. **11**(1), 10–18 (2009)
8. Han, J.: Data Mining: Concepts and Techniques. Morgan Kaufmann, USA (2005)
9. Kleppe, A.: Software Language Engineering: Creating Domain-Specific Languages using Metamodels. Addison-Wesley Professional, Reading (2008)
10. Kolovos, D.S., Paige, R.F., Rose, L.M., Williams, J.: Integrated model management with epsilon. In: France, R.B., Kuester, J.M., Bordbar, B., Paige, R.F. (eds.) ECMFA 2011. LNCS, vol. 6698, pp. 391–392. Springer, Heidelberg (2011)
11. Kühne, T.: Matters of (meta-)modeling. Softw. Syst. Model. **5**(4), 369–385 (2006)
12. Mernik, M., Heering, J., Sloane, A.M.: When and how to develop domain-specific languages. ACM Comput. Surv. **37**(4), 316–344 (2005)
13. Pelleg, D., Moore, A.: X-means: extending K-means with efficient estimation of the number of clusters. In: Proceedings of the 17th International Conference on Machine Learning, pp. 727–734. Morgan Kaufmann (2000)

14. Quinla, J.R.: C4.5: Programs for Machine Learning. Morgan Kaufmann Publishers Inc., San Francisco (1993)
15. Rice, W.: Moodle E-Learning Course Development. Packt Publishing, Birmingham (2006)
16. Rice, W.: Blackboard Essentials for Teachers. Packt Publishing, Birmingham (2012)
17. Romero, C., Ventura, S.: Data mining in education. Wiley Interdisc. Rev.: Data Mining Knowl. Discov. **3**(1), 12–27 (2013)
18. Sierra, J.L.: Language-driven software development (invited talk). In: Pereira, M.J.V., Leal, J.P., Simões, A. (eds.) 3rd Symposium on Languages, Applications and Technologies. OpenAccess Series in Informatics (OASIcs), vol. 38, pp. 3–12 (2014)
19. Steinberg, D., Budinsky, F., Paternostro, M., Merks, E.: EMF: Eclipse Modeling Framework, vol. 2. Addison-Wesley Professional, Reading (2008)
20. Wrembel, R., Koncilia, C.: Data Warehouses and Olap: Concepts, Architectures and Solutions. IRM Press, London (2006)
21. Zorrilla, M., García-Saiz, D.: A service-oriented architecture to provide data mining services for non-expert data miners. Decis. Support Syst. **55**(1), 399–411 (2013)

WSDLUD: A Metric to Measure the Understanding Degree of WSDL Descriptions

Mario Marcelo Berón[1]([⊠]), Hernán Bernardis[1], Enrique Alfredo Miranda[1],
Daniel Edgardo Riesco[1], Maria João Varanda Pereira[2],
and Pedro Rangel Henriques[3]

[1] Department of Computer Science,
Universidad Nacional de San Luis, San Luis, Argentina
{mberon,hbernardis,eamiranda,driesco}@unsl.edu.ar
[2] Centro Algoritmi, Universidade do Minho,
Instituto Politécnico de Bragança, Bragança, Portugal
mjoao@ipb.pt
[3] Department of Computer Science/Centro Algoritmi,
University of Minho, Campus de Gualtar, Braga, Portugal
prh@di.uminho.pt

Abstract. In this article, WSDL Understanding Degree (WSDLUD) a metric aimed at measuring a priori the understandability of WSDL (Web Services Description Language) descriptions is presented. In order to compute WSDLUD, all the static information available in a WSDL description is collected. This information is submitted to an evaluation process based on a method named LSP (Logic Scoring of Preference). This evaluation process outputs a Global Preference value that indicates the satisfaction level of the WSDL description regarding the evaluation focus, in this case, the understanding degree.

Keywords: WSDL · Web services comprehension · LSP

1 Introduction

Nowadays the Web Services (WS) are fundamental software artifacts for building service oriented applications. According to World Wide Web Consortium (W3C, for details see http://www.w3.org/), a WS is: *a software application identified by a URI, whose interfaces and bindings are capable of being defined, described, and discovered as XML artifacts. A WS supports direct interactions with other software agents using XML-based messages exchanged via Internet-based protocols.* The organizations, increasingly, produce web services which are used by other organizations to produce new software systems aimed at solving business demands. Web services have associated a description which specifies the data types used, the operations provided, inputs and output, the technology used to accomplish the communications between other high level and low level

© Springer International Publishing Switzerland 2015
J.-L. Sierra-Rodríguez et al. (Eds.): SLATE 2015, CCIS 563, pp. 91–100, 2015.
DOI: 10.1007/978-3-319-27653-3_9

of software elements. These descriptions are published in the internet and the organizations can retrieve them and decide if some of those services are useful for building the software they need [12]. Web Services are software packages and therefore they must be comprehend for maintenance tasks (bug fixing, adaptation, evolution, etc.). The primary information source to accomplish this task is the respective WSDL (Web Service Description Language, http://www.w3.org/TR/wsdl20/) description. Although, there are several resources from which it is possible to collect information about the Web Service, the WSDL description is the first that the user employs for analysing its usefulness for his purposes. Furthermore, the web service descriptions are interesting because they provide a high level abstraction data which can be very useful to simplify the understanding of the web services. As said above, a standard language used to write web service descriptions is WSDL. This language is a dialect of XML with well defined rules to specify each component. Being a XML based language it is fastidious to read such a description, and therefore a tool is needed to assist the software engineer in this task. In this context, many tools can be found that are oriented to facilitate the inspection of WSDL descriptions, transform to a different WSDL version, compute several metrics, produce user-friendly visualizations, etc. However, at the best of our knowledge, only a few are oriented to help their understanding. Taking this into consideration, in this article WSDLUD (Web Service Understanding Degree) is presented. WSDLUD is a metric aimed at providing, a priori, a measurement about the WSDL description understanding complexity. For calculating WSDLUD, Logic Scoring of Preference Method (LSP) [7,14] is used. LSP is a multi-criteria evaluation method; it requires a Criteria Tree, an Aggregation Structure and a set of Elementary Criteria Functions to be defined. Combining systematically such elements, this method produces a satisfaction level that indicates, in this case, the understanding degree of a WSDL description. In order to apply LSP and compute WSDLUD, the WSDL description must be statically analysed and all the information available must be retrieved. This information is submitted to different evaluation procedures in order to obtain satisfaction values (values in [0,1] or [0,100]). To perform these processes, the use of both compilation and natural language processing techniques are required. The first is used to retrieve formal elements from WSDL source code. The second is employed to gather semantic information from unstructured information sources.

The article is organized as follow. Section 2 describes the work tightly related with the research topics here presented. Section 3 defines the WSDLUD evaluation structures. Section 4 presents the case studies where it is possible to observe the results obtained through the application of WSDLUD to some test cases available in W3C. Section 5 closes the paper with some conclusions and future work.

2   Related Work

The WSDL description analysis is based on static and behavioral information. The traditional approaches are oriented to compute metrics to compare and evaluate a set of program parameters [2,13,15].

Considering static information, authors [13] have defined metrics for organization security. In this context, the authors affirm that the easier to understand a WSDL description the easier will be to carry out fraudulent actions against the organization. On account of that, the authors compute the understanding level of WSDL description and if it is high they define approaches to diminish its readability.

The second, based on behavioral information, is concerned with measuring the WSDL description considering the complexity of the operations and messages involved. The more complex the operations and messages are, the more complex will be to understand the WSDL description [9].

It is also possible to find works that use ad-hoc approaches. They are based on traditional object oriented metrics to measure quality attributes of WSDL descriptions [5,6].

WSDLUD metric, defined in this article, is different from those found in the literature in several aspects. First, all formal elements of the WSDL description (types, port types, bindings, services) are considered and for each one of them the understanding degree is measured.

Second, the WSDL description's understandability can be simplified if the informal information (those provided by the identifiers and documentation) gives useful semantic information about the description's domain. For this reason, several metrics to measure the quality of the identifiers and documentation of the description, are defined and calculated.

Third, the value of our metric is produced by the combination of other metrics (those mentioned before) which consider both formal and informal information. We used these metrics to measure WSDL descriptions and obtain a final value for each of them. This final value is computed by using a multi criteria method. This method is parameterizable allowing to reflect the engineer experience in the evaluation mechanism. Finally, as a side effect, the process used to compute WSDLUD can also be used for: (i) To provide a ranking of WSDL descriptions understandability, (ii) To build visualizations based in charts, and allow to analyse the results and to discover the possibilities to improve the WSDL description understanding.

To finish this section, it is important to notice that, at best of our knowledge, a metric with the characteristics mentioned above was not described in the literature. So, we believe that the work here reported is a valid contribution for the comprehension of WSDL specifications.

3 WSDLUD

In this section, all the concepts and processes involved in the definition and measurement of WSDLUD are described in detail.

3.1 WSDL Description Criteria Tree

The criteria tree of a WSDL description (these characteristics were extracted from a WSDL specification provided by W3C.) is composed by the following characteristics: (i) Type Understanding Degree, (ii) Message Understanding Degree, (iii) Port Type Understanding Degree, (iv) Binding Understanding Degree and (v) Service Understanding Degree. Each characteristic has an associated sub criteria tree which takes into consideration the proper properties of the evaluated element.

In the next paragraphs the characteristics mentioned above will be developed, for each of them, the Criteria Tree will be explained.

Type Understanding Degree. This characteristic is composed by the following attributes: *Number of Primitive Types, Number of Complex Types, Documentation Quality, Type Name Quality* and *Number of Fields*. Clearly, a primitive type (a primitive type is a type provided by the language), for example: text, integer, real, boolean, etc. will be easier to understand than a complex type (a complex type is a type defined by the user). A primitive type can be deduced from its identifier and the explanations provided by the language manual. A complex type is more difficult of perceiving because it is composed by several identifiers, which are susceptible to do many analysis and the explanations exposed in the language manual are not enough. In this context, if the documentation provided is bad or null, the comprehension will be even more difficult.

Message Understanding Degree. This characteristic can be evaluated taking into consideration the following attributes: *Message Documentation Quality, Message Name Quality* and *Part Understanding Degree*. Concerning the first two elements, it is possible to say that they will provide relevant information when some semantic information can be extracted. For that the following components are considered: name, element name and type. The sub-characteristic named *Part Understanding Degree* which can be divided in *Part Name Quality, Part Element Name Quality* and *Part Type Understanding Degree* attributes. All these attributes must also be considered when the message understandability needs to be measured.

Port Type Understanding Degree. This characteristic has the following attributes: *Port Type Name Quality, Port Type Documentation Quality* and *Port Operation Understanding Degree*.

The first two are important because they provide semantic information when they are well defined. Semantic information can also be extracted from *Port Operation Understanding Degree* measuring the *Port Type Operation Understanding Degree*.

The definition of this characteristic follows the same approach that message part, in other words to each simple operation element we consider some attributes like name, documentation, parameters, etc. (more details about the disaggregation of this sub-characteristic can be found in [3]).

Binding Understanding Degree. This characteristic is composed by the following attributes: *Binding Name Quality, Binding Documentation Quality, Binding Type Understanding Degree* and *Binding Operation Understanding Degree*.

Once more the name quality and the documentation quality are important characteristics to measure using the attributes: *Binding Name Quality* and *Binding Documentation Quality*. The other two attributes are already defined in others characteristics. *Binding Type Understanding Degree* is defined in *Type Understanding Degree* and *Binding Operation Understanding Degree* is defined in *Port Type Understanding Degree*. For this reason, during evaluation process we re-use the values obtained in previous computation.

Service Understanding Degree. A service is made available by a WSDL description. A service has a name and documentation and it is composed by ports. For analyzing the *Service Understanding Degree* it is necessary to measure *Service Name Quality, Service Documentation Quality* and *Service Port Understanding Degree* in a Service context.

3.2   Aggregation Structure

As LSP method states [14], the satisfaction values that result from the application of the Elementary Criteria Functions to the measurable attributes, must be aggregated in order to obtain the Global Preference. This Global Preference represents the satisfaction of the object under evaluation. As could be seen in Subsect. 3.1, we propose a Criteria Tree for each WSDL element (type, message, port, etc.). For each of these Criteria Trees, we developed a specific Aggregation Structure. To illustrate the approach and to save space, in Fig. 1 we only show the Aggregation Structure for the characteristic *Message Understanding Degree*.

We used a partial absorption LSP function (compound by operator A (arithmetic mean) and SQU (square mean) — all the LSP operators are better explained in [8]) to aggregate Message Documentation Quality and Message Name Quality. This kind of asymetric compound operators are used when some input values could be zero (non-mandatory input). It is necessary because in many cases, messages do not have a good documentation (sometimes do not have at all). A medium conjunctive operator (CA) is used to compute the Message Understanding Degree Global Preference. This kind of operator is employed when the input requirements are mandatory. Thus if one of the input values is zero, the operation result will be zero. The weights are used to express the relative importance of input preference. As message documentation and name provides more significant semantic information, its weight is 70 %, as opposed to *Part Understanding Degree* which provides less semantic information (its weight is 30 %).

3.3   Information Extraction Techniques and Elementary Criteria Functions

The information extraction techniques and the Elementary Criteria Functions are the most important features for the evaluation process that will be described.

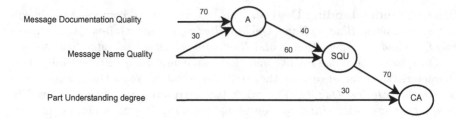

Fig. 1. Message understanding degree Aggregation Structure.

The former allows to obtain the information and perform all the analysis to get each attribute value for the Criteria Tree. The latter maps each of these in a satisfaction level, i.e., a value in the interval [0,1] (or [0,100]). This value represents the satisfaction degree of the attribute for the object under evaluation according to the sensibility and experience of the authors.

Information Extraction Techniques. The approach used to extract information from a WSDL description combines compilation techniques, natural language processing algorithms and strategies to compute indicators [4]. The first are implemented using DOM (Domain Object Model) a parser for XML language which explicitly builds an internal representation of the analysed XML source code. Several traversals are applied through this internal representation for gathering the desired information. The identifiers and the documentation are extracted by using compilation techniques. In order to retrieve semantic information IdA (Identifier Analysis) [1] is used. IdA is a tool aimed at applying algorithms to divide, expand and find a meaning for the identifiers of a program. Finally, with the goal to provide a measure about of the understanding degree of a WSDL description, NESSy [11] was used. NESSy is a tool to evaluate software based on LSP method.

For attributes like *Type Name Quality* (see in Algorithm 1 the computation process of the satisfaction level of Type Name Quality Criterion), *Message Name Quality* or *Binding Name Quality* we use identifier analysis techniques.

The purpose of this analysis is to discover the relation between the names and the concepts of the problem domain. The name quality is higher when its related words are meaningful. The result of the techniques is a percentage which indicates the satisfaction level for a particular name quality.

For attributes like *Type Documentation Quality, Message Documentation Quality, Binding Documentation Quality*, etc., we use documentation analysis techniques. This kind of attributes has as main goal to measure the usefulness level of the information provided by the element's documentation (IdA also is used to carry out this task). The analysis techniques gathers documentation and returns a percentage which represents the satisfaction level for the attribute under study. In first place the documentation is divided by words, then the irrelevant words are deleted. The next step consists of analysing each word and count those that have a useful meaning. The result is obtained carrying out the following computation: $\frac{Number\ of\ Word\ with\ Mean}{Number\ of\ Words}$.

Algorithm 1. Satisfaction Level of Type Name Quality Criterion

---

input : *typeName* a string which represents a type name.
output: Satisfaction Level, a percentage that indicates the
        satisfaction level of the criterion Type Name
        Quality.
Data: *wordSet,stopWords* a set of words.
Data: *pal* a string which represent a word extracted from a
        type name.
Data: *wordsWithMeans* an integer variable which counts the
        number of words extracted from a type name which have
        meaning.
wordSet←division(typeName);
stopWords←extractStopWords(wordSet);
wordSet←wordSet-stopWords;
wordsWithMean←0;
foreach *w in wordSet* **do**
        pal←expand(w);
        if *hasMean(pal)* **then**
            | wordsWithMean←wordsWithMeans + 1;
end
return ($\frac{wordsWithMeans}{|wordSet|}$);

Elementary Criteria Functions. In this evaluation process, the majority of Elementary Criterion Function are direct mappings, since most of the attributes values are computed by extraction techniques. They take as input the strings to be analysed and return a percentage value that could directly be mapped to a satisfaction value.

4   Case Study

This section presents the evaluation of five WSDL descriptions using LSP and the structures defined in Sect. 3 [10]. All descriptions belong to web services frequently used by information systems:

(i) *Google Web APIs* (https://code.google.com/p/dic/downloads/detail? name=GoogleSearch.wsdl), provides operations to do Google searchs, (ii) *Create Queue (Amazon)* (http://queue.amazonaws.com/doc/2009-02-01/ QueueService.wsdl), offers a reliable, highly scalable hosted queue for storing messages as they travel between computers, (iii) *Airport* (http://www. webservicex.com/airport.asmx?wsdl), provides useful information of all world airports (e.g. airport codes, names, countries, countries code, latitude, longitude, etc.) (iv) *Global Weather* (http://wsf.cdyne.com/WeatherWS/Weather. asmx?WSDL), gets weather report for all major cities around the world, and

(v) *OFAC* (http://www.webservicex.net/OFACSDN.asmx?WSDL) aids banks in meeting the requirements of the US Treasury Department's Office of Foreign Asset Control (OFAC).

Table 1. Partial and global evaluation of WSDL

High-Level characteristic	Google	Weather	Amazon	Airport	OFAC
Types U. D	60,2665	71,5131	68,8148	72,2303	40,5846
Messages U. D	69,1173	83,3624	79,753	77,4924	58,8801
Port Types U. D	75,7194	81,4166	82,1289	81,8902	45,3519
Bindings U. D	75,5258	79,3457	82,2505	79,5241	42,755
Services U. D	78,9946	79,6724	89,4138	79,7011	42,0794
Final Scores	71,5594	77,0112	80,1496	78,0884	45,4495

Table 1 shows the global understanding degree for each WSDL description. Each *Global Preference* was computed aggregating all the characteristic preferences with the logical operator *CA* (this function simulates simultaneity) and the weight equally distributed among the characteristics (20 % for each one). The choice of this operator is due to the fact that all WSDL components (type, message, port type, etc.) must be understandable. If one of these is incomprehensible, the whole WSDL will be difficult to understand.

As can be seen in Table 1, almost all WSDL are very similar taking into account understanding degree, except for *OFAC* WSDL description. This is because that description has numerous identifiers with acronyms which decreases the satisfaction levels.

Weather and *Airport* define each type using a few primitive and complex types. Furthermore they specify explicit and unambiguos identifiers. On the other hand, *Google* uses a number of primitive and complex types that exceed the established thresholds. The majority of messages's parts of *Weather* WSDL uses primitive types and this fact rise its *Messages Understanding Degree* satisfaction value.

In general, *Amazon* WSLD presents more documentation than others in different parts, like messages, types, port types and services. This makes this WSDL the most understandable of the case study.

From another point of view, this set of metrics was proposed to measure each component individually inside a WSDL. In this sense, we could compare, for example, all elements of a kind that a WSDL contains (e.g. types, messages or services), in order to analyze it individually. This is could be useful for maintainability or re-structuring purposes. In this context, we measure three messages that presents *Create Queue (Amazon)* WSDL description. In this context, we measure the quality of three different messages of the Create Queue (Amazon) WSDL description and the results can be seen in Table 2.

As can be seen, *RemovePermissionRequest* (RPR) message is the most understandable of these three messages and *SendMessageResponse* (SMR) the worst. This is basically due to *Message Part Understanding Degree* satisfaction values.

Table 2. Messages individual measurement of *Create Queue (Amazon)* WSDL description.

Sub-characteristic	SendMessage Response	RemovePermission Request	DeleteMessage Response
M. Doc. Quality	0	0	0
M. Name Quality	100	100	100
M. Parts U. D	60,9759	93,6933	73,1726
Final Scores	73,17	83,5379	77,6729

This is a comparative analyse that allows to identify the most critical parts of the description. If we want to analyse the results individually we would say that a score less than 50 % represents a candidate description for improvement.

5   Conclusion and Future Work

In this article WSDLUD a metric, to measure the understanding degree of WSDL description, was defined. In order to compute WSDLUD other metrics were also specified. These metrics have as main goal to provide an estimation about the understanding degree of each description part. Each part is associated with an importance level specified by the engineer. Both values (understanding degree and importance level) are used by LSP (a multi criteria evaluation method) to produce a global value which represents the desired WSDL description understanding degree.

We believe that our approach is novel because it makes possible to analyse each part of a particular WSDL description as well as the global understanding degree. Yet more important, all the engineer's experience can be included in the evaluation process in order to get more significant results. All the detailed information provided by our system can be used to identify the most critical parts of the description and the chances for quality improvement. In some cases, the description can be simplified or made more readable. But, in other cases, the complexity of the description is full dependent on the domain complexity and there is not chance for improvement.

As future work we intend to:

(i) Improve the Criteria Tree (CT) and Aggregation Structure (AS);
(ii) Extend the work presented in this paper to WSDL 2.0;
(iii) Apply a similar analysis to study business processes specified with BPEL (Business Process Execution Language).

Acknowledgements. This work has been supported by FCT–Fundação para a Ciência e Tecnologia within the Project Scope: UID/CEC/00319/2013.

References

1. Azcurra, J., Berón, M., Montenjano, G., Farnese, A., Henriques, P., Pereira, M.: AId: Uma Ferramenta para Análise de Identificadores de Programas Java. In: Congreso Nacional de Ingeniería Informática/Sistemas de Información, pp. 880–892, Noviembre 2014
2. Bernardis, H., Beron, M., Riesco, D., Henriques, P.R.: Extracción de información y cálculo de métricas en WSDL 1.1 y 2.0. In: Congreso Nacional de Ingeniería Informática/Sistemas de Información, pp. 963–974, Noviembre 2014
3. Beron, M., Henriques, P.R., Riesco, D., Pereira, M.J.V.: On the Comprehension of WSBPEL Programs. Technical report, Universidad Nacional de San Luis - Universidade do Minho (2015)
4. Carvalho, N.R.: An Ontology Toolkit for Problem Domain Concept Loction in Program Comprehension. Ph.D. thesis, Escola de Engenaria, Universidade do Minho (2014)
5. Coscia, L.O., Crasso, M., Mateos, C., Zunino, A.: Estimating Web Service interface quality through conventional object-oriented metrics. CLEI Electron. J. **16**(1) (2013)
6. Coscia, L.O., Mateos, C., Crasso, M., Zunino, A.: Refactoring code-first Web Services for early avoiding WSDL anti-patterns: approach and comprehensive assessment. Sci. Comput. Program. **89**, 374–407 (2014)
7. Dujmovic, J.: Continuous preference logic for system evaluation. IEEE Trans. Fuzzy Syst. **15**(6), 1082–1099 (2007)
8. Dujmovic, J.: Characteristic forms of generalized conjunction/disjunction. In: IEEE International Conference on Fuzzy Systems, FUZZ-IEEE 2008, (IEEE World Congress on Computational Intelligence), pp. 1075–1080. IEEE (2008)
9. Kumar, R., Indraveni, K., Goel, A.K.: Automation of detection of security vulnerabilities in Web Services using dynamic analysis. In: 9th International Conference on Internet Technology and Secured Transactions (ICITST), pp. 334–336, December 2014
10. Liu, L., Sun, T., Fang, W., Liu, N.: Usability evaluation of the subway train dispatching system. In: 2011 International Conference on Information Science and Technology (ICIST), pp. 1123–1128, March 2011
11. Miranda, E., Berón, M., Montejano, G., Pereira, M.J.V., Henriques, P.R.: NESSy: a New Evaluator for Software Development Tools. In: 2nd Symposium on Languages, Applications and Technologies, SLATE 2013, Porto, Portugal, pp. 21–37, 20–21 June 2013
12. Newcomer, E.: Understanding Web Services: XML, WSDL, SOAP, and UDDI. Addison-Wesley Professional, New York (2002)
13. Sripairojthikoon, P., Senivongse, T.: Concept-based readability measurement and adjustment for web services descriptions. In: 16th International Conference on Advanced Communication Technology (ICACT), pp. 378–388, February 2014
14. Su, S., Dujmovic, J., Batory, D.S., Navathe, S.B., Elnicki, R.: A cost-benefit decision model: analysis, comparison and selection of data management. ACM Trans. Database Syst. **12**(3), 472–520 (1987)
15. Tibermacine, O., Tibermacine, C., Cherif, F.: A Practical Approach to the Measurement of Similarity between WSDL-basedWeb Services. RNTI: Revue des Nouvelles Technologies de l'Information, Special Issue CAL 2013 (RNTI-L-7): 03–18 (2014)

Combining Processing with Racket

Hugo Correia and António Menezes Leitão[✉]

INESC-ID, Instituto Superior Técnico,
Universidade de Lisboa, Rua Alves Redol 9, Lisboa, Portugal
{hugo.f.correia,antonio.menezes.leitao}@tecnico.ulisboa.pt

Abstract. Processing is a programming language created to teach programming in a visual context. Despite its success, Processing remains a niche language with limited applicability in the architectural field, as no Computer-Aided Design (CAD) application supports Processing. This work presents an implementation of Processing for the Racket platform, that transforms Processing code into semantically equivalent Racket code. Our Processing implementation is developed as a Racket module language for interoperability with Racket and other module languages of Racket's language ecosystem. Our implementation allows us to take advantage of Rosetta, a Racket library that provides access to several CAD back-ends (e.g. AutoCAD, Rhinoceros, SketchUp). As a result, architects and designers can take advantage of our implementation to use Processing with their favourite CAD application.

Keywords: Processing · Racket · Compilers · Interoperability

1 Introduction

Processing [1] is a programming language and development environment created to teach programming in a visual context. The language has grown over the years, creating a community where users can share their artistic works. Many examples and educational materials are available to newcomers, reducing their effort to learn the language. Moreover, Processing offers a wide range of 2D and 3D drawing primitives, as well as an Integrated Development Environment (IDE) that provides tools to programmatically create innovative designs.

Nonetheless, Processing is a niche programming language with limited applicability in the architectural field, as architects depend on traditional heavyweight CAD applications (e.g. AutoCAD, Rhinoceros 3D, etc.), that provide APIs tailored for that specific CAD tool. Unfortunately, no CAD application allows users to write scripts in Processing. As a result, architects that have learnt Processing cannot use the language or any of the publicly available examples to program in the context of their favourite CAD tool. This problem is addressed in this paper, showing how our solution combines Processing with the Racket programming language.

Racket [2] is a descendent of Scheme, which encourages developers to tailor their environment to project-specific needs, offering an ecosystem that allows

© Springer International Publishing Switzerland 2015
J.-L. Sierra-Rodríguez et al. (Eds.): SLATE 2015, CCIS 563, pp. 101–112, 2015.
DOI: 10.1007/978-3-319-27653-3_10

the creation of new languages which have direct interoperability with other Racket libraries. For instance, Rosetta [3] is a Generative Design tool built on top of Racket, that encompasses Racket's philosophy of using different languages. Rosetta allows programmers to generate 2D and 3D geometry in a variety of CAD applications, namely AutoCAD, Rhinoceros3D, SketchUp, and Revit, using several programming languages, such as JavaScript, AutoLISP, Racket, and Python. Furthermore, Racket offers a pedagogic IDE, DrRacket, which can be adapted to support new module languages of the Racket ecosystem.

Our solution is to implement a source-to-source compiler that translates Processing code to semantically equivalent Racket code, enabling architects to prototype designs using Processing in a CAD tool. Moreover, as Racket encourages developers to use and create different languages within the Racket ecosystem [4], we have developed an interoperability mechanism to access Racket libraries and to combine Processing with scripts written in other languages of the Racket ecosystem, such as Python [5] or Typed Racket [6].

2   Processing

Processing was developed at MIT media labs and was heavily inspired by the *Design by Numbers* [7] project. The language was created to teach computer science to artists and designers with no previous programming experience. Processing has grown over the years with the support of a large community, which has written several educational materials, demonstrating how programming can be used in the visual arts.

Processing can be considered a dialect of the Java programming language, that significantly simplifies the original language. For instance, in Java, developers have to implement a large set of steps to develop simple examples, namely a public `class` that implements public methods and a static `main` method. These constructs bring an initial overhead and verbosity for novice programmers, which are cumbersome for beginners that want to quickly try out new ideas. To solve this problem, Processing allows users to write simple scripts (i.e. simple sequences of statements) that do not have the verbosity of Java, thus enabling them to quickly create new designs.

The Processing language introduces the notion of a *sketch*, which is used to organize source code. A *sketch* can operate in one of two distinct modes: *Static* or *Active*. *Static mode* supports simple Processing scripts, such as simple statements and expressions. *Active mode* allows users to implement their *sketches* using more advanced features of the language. If a function or method definition is present, the *sketch* is considered to be in *Active mode*. Within each *sketch*, Processing users can define two functions to aid their design process: `setup` and `draw`. On one hand, the `setup` function is called once when the program starts. In `setup` the user can define initial environment properties and execute initialization routines needed to create the design. On the other hand, the `draw` function runs after `setup` and executes the code to draw the design. The control flow is simple: first `setup` is executed, setting-up the environment; followed by `draw` called in a loop, rendering the sketch until it is stopped by the user.

Moreover, Processing offers users a set of tools that are specially tailored for visual artists. For instance, 2D and 3D drawing primitives are made available, rendering designs in different 2D and 3D rendering environments. Processing also offers a simple but effective development environment called the PDE (Processing Development Environment), where users can develop their programs using a tabbed editor with IDE services such as syntax highlighting and automatic code formatting.

3 Related Work

The following section presents different approaches that influenced our work, and an analysis of their main features.

3.1 Processing.js

Processing.js [8] is a JavaScript implementation of Processing for the Web that enables developers to create scripts in Processing or JavaScript. Using Processing.js, developers can use Processing's approach to design 2D and 3D geometry in a HTML5 compatible browser. Processing.js uses a custom-purpose JavaScript parser, that parses both Processing and JavaScript code, translating Processing code to JavaScript while leaving JavaScript code unmodified. Moreover, Processing.js implements Processing drawing primitives and built-in classes directly in JavaScript, allowing for greater interoperability between both languages, as Processing code is seamlessly integrated with JavaScript. To render Processing scripts in a browser, Processing.js uses the HTML canvas element to provide 2D geometry, and WebGL to implement 3D geometry. Processing.js encourages users to develop their scripts in Processing's development environment, and then render them in a web browser. Additionally, Sketchpad [9] is an alternative online IDE for Processing.js, where users can create and test their design ideas online and share them with the community.

3.2 Processing.py and Ruby-Processing

Ruby-Processing [10] and Processing.py [11] produce Processing/Java as target code. Both Ruby and Python have language implementations for the JVM, allowing them to directly use Processing's drawing primitives. Processing.py takes advantage of Jython to translate Python code to Java, while Ruby-Processing uses JRuby to provide a Ruby wrapper for Processing. Processing.py is fully integrated within Processing's development environment as a language mode, and therefore provides an identical development experience to users. On the other hand, Ruby-Processing is lacking in this aspect, by not having a custom IDE. However, Ruby-Processing offers *sketch* watching (code is automatically executed when new changes are saved) and live coding, which are functionalities that are not present in any other implementation.

3.3 ProfessorJ

ProfessorJ [12,13] was developed to be a language extension for DrScheme [14]. ProfessorJ implements a traditional compiler pipeline, that starts with lexing and parsing phases, producing an intermediate representation in Scheme. Subsequently, the translated code is analysed, generating target Scheme code by using custom defined functions and macro transformations. ProfessorJ implements several strategies to map Java code to Scheme. For instance, Java classes are translated into Scheme classes with certain caveats, such as implementing static methods as Scheme procedures or by changing Scheme's object creation to appropriately handle Java constructors. Also, Java has multiple namespaces while Scheme has a single namespace, hence name mangling techniques were implemented to correctly support multiple namespaces. Moreover, Java's built-in primitive types and some classes are directly implemented in Scheme, while remaining classes are implemented in Java. Strings, Arrays, and Exceptions are mapped directly into Scheme forms. Implementing them in Scheme is possible (with some constraints) due to similarities in both languages which, in turn, allow for a high level of interoperability. Finally, ProfessorJ is fully integrated with DrScheme, providing a development environment that offers syntax highlighting, syntax checking, and error highlighting for Java code.

4 Compilation Process

Observing all previous implementations, it is clear that an IDE is an important feature to have in any implementation of the Processing language (as only Ruby-Processing is lacking one). Regarding the runtime system, Processing.js and ProfessorJ implement it in the target language to achieve greater interoperability; while, Ruby-Processing and Processing.py take advantage of JVM language implementations to provide Processing's runtime.

Finally, we observe that none of the presented approaches offers a solution that allow us to explore Processing in the context of a CAD environment. Neither Processing.js, Processing.py, or Ruby-Processing allow designs to be visualized in a CAD tool. Alternatively, other external Processing libraries could be explored to connect Processing with CAD applications. For instance, OBJExport [15] is a Processing library to export coloured meshes from Processing as OBJ or X3D files. These files can then be imported into some CAD applications. However, using this approach, we lose the interactivity of programming directly in a CAD application, as users have to generate and import the OBJ file each time the Processing script is changed, creating a cumbersome workflow. Moreover, as shapes are transformed to meshes of triangles and points, there is a considerable loss of information, as the semantic notion of the shapes is lost.

Our proposed solution was to develop Processing as a new Racket language module, using Rosetta for Processing's visual needs, and integrating Processing with DrRacket's IDE services. We chose Racket, firstly, because it simplifies the development of new languages, providing libraries to implement the lexical and syntactic definitions of the Processing language, as well as offering mechanisms to generate semantically equivalent Processing code. Secondly, Racket's capabilities enable us to easily adapt our Processing implementation to work with DrRacket (Racket's educational IDE), providing an IDE to its users. Moreover, after analysing ProfessorJ, we concluded that many parts of the lexical and syntactical definitions, and type-checking procedures could be adapted, due to the similarities between Java's and Processing's language definitions. Finally, our implementation allows us to take advantage of Rosetta to augment Processing with capabilities that make the language suitable for architectural work.

Our Processing implementation follows the traditional compiler pipeline approach (illustrated in Fig. 1), composed by three separated phases, namely parsing, code analysis, and code generation.

4.1 Parsing Phase

The compilation process starts with the parsing phase, which is divided in two main steps. First, Processing source code is read and transformed into tokens. Secondly, tokens are given to an LALR parser, building an abstract syntax tree (AST) of Racket objects which will be analysed in subsequent phases. To implement the lexer and parser specifications, we used Racket's **parser-tools** [16] library, adapting parts of ProfessorJ's lexer and grammar specification to fit Processing's needs.

4.2 Code Analysis

Following the parsing phase, an analysis of the AST must be made, due to differences between Processing's and Racket's language definitions. For instance, Processing has static type-checking and different namespaces for methods, fields, and classes, while Racket is dynamically typed and has a single namespace. As a result, custom tailored mechanisms were needed to correctly type-check the AST and support Processing's scoping rules.

Firstly, the AST is traversed passing scope information to child nodes. When a new definition is created, be it a function, variable, or class, the newly defined binding is added to the current node's scope along with its type information. Each time a new scope is created in Processing, a new custom scope is created to represent it, referring to the current scope as its parent. These mechanisms are needed to implement Processing scoping and type-checking rules. For example, the information of the return type, arity, and argument types are needed to type-check a function call.

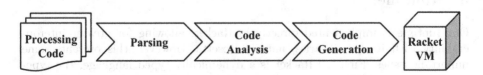

Fig. 1. Overall compilation pipeline

Secondly, the type-checking procedure runs over the AST starting topmost AST node. As before, it repeatedly calls the type-checker on child nodes until the full AST is traversed, using previously saved bindings in the current scope to find out the types of each binding. During the type-checking procedures, each node is tested for type correctness and, in some cases, promoting types if necessary. In the event that types do not match, a type error is produced, signalling where the type error occurred.

4.3   Code Generation

After the AST is fully analysed and type-checked, semantically equivalent Racket code can be generated. To achieve this, every AST node generates Racket code by using custom defined macros and functions. Afterwards, Racket will expand the defined macros and load the generated code into Racket's VM. By using macros we can create boilerplate Racket code that can be constantly modified and tested by the developer

Racket and Processing follow the same evaluation order on their programs, thus most of Processing's statements and expressions are directly mapped into Racket forms. However, other statements such as `return`, `break`, or `continue` need a different handling, as they use control flow jumps. To implement this behaviour, we used Racket's *escape continuations* [17] in the form of `let/ec`.

Furthermore, Processing has multiple namespaces, which required an additional effort to translate bindings to Racket's single namespace. To support multiple namespaces in Racket, binding names were mangled with custom tags. For instance, a `fn` tag is appended to functions, so function `foo` internally would be `foo-fn`. The use of '-' as a separator allows us to solve the problem of name clashing with user defined bindings, as Processing does not allow '-' in names. Also, as we have function overloading in Processing, we append specific tags that represent the argument's types to the function's name. For instance, the following function definition: `int foo(float x, float y){ ... }` would be translated to (`define (foo-FF-fn x y) ...`).

To correctly support Processing's distinctions between *Active* and *Static* mode, we used the following strategy. We added a custom check in the parser that signals if the code is in *Active mode*, i.e. if a function or method is defined. In this mode, global statements are restricted, thus when generating code for global statements we check if the code is in *Active mode*, if so we signal an error indicating the invalid statement.

5   Runtime

Our runtime is implemented directly in Racket, allowing for greater interoperability with Racket libraries, namely Rosetta. However, this presents some important issues. First, as Racket is a dynamically typed language, the type-checker, at compile time, cannot know the types of Racket bindings. To solve this problem, we introduced a new type in the type hierarchy, which the type-checker ignores when type checking these bindings. Furthermore, as Processing primitives and built-in classes are implemented in Racket, we also have the problem of associating type information for these bindings. Therefore, we created a simple macro that allows us to associate type information to Racket definitions, by adding them to the global environment, thus the type-checker can correctly verify if types are compatible.

Moreover, Processing's drawing paradigm closely resembles OpenGL's traditional push & pop matrix style. To provide rendering capabilities in our system, we use Rosetta, as it provides design abstractions that not only let us generate designs in an OpenGL render, but also give us access to several CAD back-ends. Custom interface adjustments are needed to implement Processing's drawing primitives in Racket, as not every Processing primitive maps directly into Rosetta's. Furthermore, Rosetta also enables us to provide with additional drawing primitives that are unavailable in the original Processing environment.

6   Interoperability

One of the advantages of developing a source-to-source compiler is the possibility of combining libraries that are written in different languages. The Racket platform encourages the use and development of different languages to fulfil programmers' needs, offering a set of extension mechanisms that can be applied to many of the language's features. The combination of Racket's language modules [6] and powerful hygienic macro system [18] enables users to extend the base Racket environment with new syntax and semantics that can be easily composed with modules written in different dialects.

To achieve interoperability with Racket, we developed Processing's compilation units as a Racket language module, adding Processing to Racket's language set. Nonetheless, compatibility issues between languages arise when accessing exported bindings from a Racket module. First, a new `require` keyword was introduced to specifically import bindings from other modules. This `require` maps directly to Racket's `require` form, receiving the location of the importing module. By using Racket's `require` we have access to all of Racket's require semantics, enabling the programmer to select, exclude, or rename imported binding from the required module.

Furthermore, Racket and Processing have different naming rules. For instance, function `foo-bar!` is a valid identifier in Racket but not in Processing, thus we cannot reference the `foo-bar!` function in our Processing code. To solve this issue, we use a translation procedure that takes a Racket identifier and transforms it into a valid Processing identifier. For example, `foo-bar!` would be translated to `fooBarBang`. Therefore, for each provided binding of a required module, we apply the translation procedure on each binding, making it available to the requiring module. By providing an automatic translation, the developer's effort is reduced, as he can quickly use any Racket module with his Processing code. Notwithstanding, as developers may not be satisfied with our

```#lang racket```
 	
```(provide foo-bar)```
```(define (foo-bar foo)```
``` ...)```
	```#lang processing```
	
	```require "foo-bar.rkt";```
	```void checkFoo(String s) {```
	``` println(fooBar(s));```
	```}```

Fig. 2. The `foo-bar` module in Racket   **Fig. 3.** `checkFoo` in Processing

automatic translation procedure, they can develop their custom mappings in a Racket module adhering to Processing identifier's rules.

Another issue that arises by importing foreign bindings, is making them accessible to our custom environment and type-checker, as they are needed during the *code analysis* phase. To solve this issue, we dynamically load the required module, saving exported bindings along with their arity. As Racket is dynamically typed, we use a special type for arguments and return types that the type-checker skips. As a result, when using bindings with this type, typing errors will only be observed when these bindings are executed at runtime. To illustrate the interoperability mechanism consider the `foo-bar` module Fig. 2, which provides the `foo-bar` function, and the Processing code illustrated in Fig. 3.

As illustrated in Fig. 3, the function `checkFoo` uses the `foo-bar` procedure from `foo-bar.rkt`. Note that our automatic translation procedure has been applied to provided bindings from the `foo-bar.rkt` modules. So in `checkFoo`, we use the automatically translated `fooBar` identifier to refer to `foo-bar`.

To understand how this is accomplished, our `require` uses a custom macro that receives the module's path (i.e. the location of the required module), as well as a list of pairs that map the original bindings of the module into their mangled form. To compute this list, we used Racket's `module->exports` primitive to provide the list of exported bindings. However, this information does not suffice, as we need to know the arity of each exported binding. This is information is needed to produce a compatible biding (i.e. a mangled binding) with our generated code. Therefore, we analysed each exported binding by `module->exports`, and retrieved its arity using the `procedure-arity` primitive. This way we can correctly perform the translation of external bindings to valid Processing identifiers and generate bindings that work with our code generation process. Lastly, when generating Racket code, our custom macro expands to Racket's `require` form, making each mangled binding available in the requiring module.

7   Example

Developing a source-to-source compiler has the advantage of allowing us to explore libraries written in another language. We provide an example of our implementation, showing how Processing code can take advantage of libraries

```
require "fib.rkt"; require "draw.rkt";
void echo(int n, Object pos, float ang, float r) {
  if ( n == 1) {
    fullArc( pos, r, ang, HALF_PI, 20);
  } else {
    fullArc( pos, r / fib(n), ang, HALF_PI, 20);
    echo(n-1, pos, ang, r);
  }
}
void mosaics(float l, int size) {
  for(int i = 0; i < size;  i++) {
    for (int j = 0; j < size; j++) {
      echo(10, xyz(i*l, j*l, 0), 0, 1);
      echo(10, xyz(i*l+1, j*l, 0), HALF_PI, 1);
      echo(10, xyz(i*l+1, j*l+1, 0), PI, 1);
      echo(10, xyz(i*l, j*l+1, 0), 3/2 * PI, 1);
    }
  }
  frame(xyz(0,0,0), size * l, 20);
}
```

Fig. 4. Processing code to generate mosaics

that were previously built for 3D modelling. This is still a work in progress, thus
the compilation results are likely to change.

Consider the Processing code presented in Fig. 4. The `mosaics` procedure
generates a grid of mosaics given the length of each mosaic and the total size
of the grid. This function uses `echo` to generate the interior pattern of each
mosaic, progressively generating smaller arcs from each corner of the mosaic.
After generating the interior pattern, the `frame` generates the full outer boundary
of the grid.

This example illustrates the use of two external Racket libraries. First, we
`require` the `fib.rkt` module to use `fib` to compute the reducing factor of
arches size. This illustrates how we can use simple Racket code with Processing.
Secondly, we `require draw.rkt`, which allows us to access `fullArc` and `frame`.
These functions enable us to generate the arcs and produce the enclosing boundary, showing how we can use previously created Racket drawing libraries with our
Processing implementation. Moreover, observe the use of `xyz` primitive. Rosetta
provides custom mechanisms to abstract coordinate systems, namely cartesian
(`xyz`), polar (`pol`), and cylindrical (`cyl`) which can be used and combined inter-
changeably. As a result, we made these abstractions (`xyz`, `pol`, and `cyl`) available
in our system, so that users can take advantage of them in their designs. Figure 5
illustrates an execution of the `mosaics` function in AutoCAD.

Observe the generated Racket code for `echo` displayed in Fig. 6. The first
point that is immediately visible is that function identifiers are renamed to support multiple namespaces. We can see that the `echo` identifier is translated to

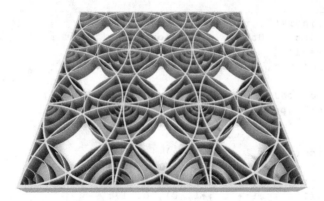

Fig. 5. Mosaics generated in AutoCAD

`echo-IOFF-fn`. Theses tags indicate the argument types of the function, where **F**, **O**, and **I**, represent the types `float`, `Object`, and `int`. Also note that imported bindings `full-arc` use the type **O** for their arguments, enabling the type-checker to correctly deal with these imported bindings. Functions and macros such as `p-div`, `p-sub`, or `p-call` are used to implement Processing's language primitives. Function are defined within a `let/ec` form, to support return semantics in functions. However, `let/ec` is not always needed and can be removed, for instance, in the case of unnecessary tail returns or when functions have return type `void`.

```
(p-function (echo-IOFF-fn n p a r)
  (let/ec return
    (p-block
    (p-if (p-eq n 1)
      (p-block (p-call fullArc-00000-fn p r ang HALF_PI h))
      (p-block (p-call fullArc-00000-fn p (p-div r (p-call
    fib-0-fn n)) a HALF_PI h)
        (p-call echo-IOFF-fn (p-sub n 1) p a r)))))))
```

Fig. 6. Generated Racket code for `echo`

We demonstrate another example (shown in Fig. 7) of our Processing implementation using libraries that are written in another language and renders designs in AutoCAD and Rhinoceros 3D. To produce this example, our Processing code `requires "elliptic-torus.rkt"`, a library written in the Racket language that is capable of generating highly parametric elliptic torus. Using this library, we can specify in Processing, the domain range, the thickness of the surface, the size of the surfaces' holes, etc.

The possibility of accessing libraries written in different languages of the Racket ecosystem enables Processing users to take advantage of the capabilities

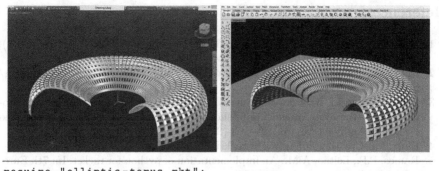

```
require "elliptic-torus.rkt";
float aMin = QUARTER_PI, aMax = 7 * aMin, h = .005;

ellipticTorus(xyz(0,0,0), h, .03, .5, aMin, aMax, 0, TWO_PI);
```

Fig. 7. Elliptic torus generated in AutoCAD and Rhinoceros 3D

of these libraries in their artistic endeavours. Moreover, these examples demonstrate that users can effortlessly migrate to our system and directly use libraries that were previously developed in Racket.

8 Conclusion

Translating a high-level language to another enables the possibility of accessing libraries that are written in different languages. Combining Processing with Racket, allows users to access libraries written in any language of the Racket ecosystem. One particularly important library is Rosetta, a portable Generative Design library that allows architects to use Processing to generate designs in a CAD application, thus providing a motivating reason for the architecture community to use our system.

Our implementation follows the common compiler pipeline architecture, generating semantically equivalent Racket code and loading it into Racket's VM. Our approach was to develop the parts of the language that Processing users most need, that empower them to write simple scripts. In future, our goal is to further develop our existing work, progressively introducing more advanced mechanisms, such as implementing Processing's class system and exception handling.

Acknowledgements. This work was partially supported by national funds through Fundação para a Ciência e a Tecnologia (FCT) with reference UID/CEC/50021/ 2013, and by the Rosetta project under contract PTDC/ATP-AQI/5224/2012.

References

1. Reas, C., Fry, B.: Processing: programming for the media arts. AI Soc. **20**(4), 526–538 (2006)
2. Flatt, M., Findler, R.B.: The racket guide (2011). http://docs.racket-lang.org/guide/. Accessed 02 May 2014
3. Lopes, J., Leitão, A.: Portable generative design for CAD applications. In: Proceedings of the 31st Annual Conference of the Association for Computer Aided Design in Architecture, pp. 196–203 (2011)
4. Flatt, M.: Creating languages in racket. Commun. ACM **55**(1), 48–56 (2012)
5. Ramos, P.P., Leitão, A.M.: An implementation of python for racket. In: 7th European Lisp Symposium, p. 72 (2014)
6. Tobin-Hochstadt, S., St-Amour, V., Culpepper, R., Flatt, M., Felleisen, M.: Languages as libraries. In: Proceedings of the 32nd ACM SIGPLAN Conference on Programming Language Design and Implementation, pp. 132–141. ACM (2011)
7. Maeda, J.: Design by Numbers. MIT Press, Cambridge (1999)
8. Resig, J., Fry, B., Reas, C.: Processing. js (2008)
9. Bader-Natal, A.: Sketchpad (2011). http://sketchpad.cc/. Accessed 28 April 2015
10. Ashkenas, J.: Ruby-processing (2015). https://github.com/jashkenas/ruby-processing. Accessed 28 April 2015
11. Feinberg, J., Gilles, J., Alkov, B.: Python for processing (2014). http://py.processing.org/. Accessed 28 April 2015
12. Gray, K.E., Flatt, M.: Compiling java to PLT scheme. In: Proceedings of 5th Workshop on Scheme and Functional Programming, pp. 53–61 (2004)
13. Gray, K.E., Flatt, M.: ProfessorJ: a gradual introduction to java through language levels. In: Companion of the 18th Annual ACM SIGPLAN Conference on Object-Oriented Programming, Systems, Languages, and Applications, pp. 170–177. ACM (2003)
14. Findler, R.B., Flanagan, C., Flatt, M., Krishnamurthi, S., Felleisen, M.: DrScheme: a pedagogic programming environment for scheme. In: Glaser, H., Hartel, P., Kuchen, H. (eds.) PLILP 1997. LNCS, vol. 1292, pp. 369–388. Springer, Heidelberg (1997)
15. Louis-Rosenberg, J.: Objexport (2013). http://n-e-r-v-o-u-s.com/tools/obj/. Accessed 29 April 2015
16. Owens, S.: Parser tools: lex and yacc-style parsing (2011). http://docs.racket-lang.org/parser-tools/. Accessed 22 September 2014
17. Flatt, M, Findler, R.B.:. The racket guide, chapter 10.3 continuations (2011). http://docs.racket-lang.org/guide/conts.html?q=continuations. Accessed 05 May 2014
18. Flatt, M.: Composable and compilable macros: you want it when? SIGPLAN Not. **37**(9), 72–83 (2002)

Batched Evaluation of Full-Sharing Multithreaded Tabling

Miguel Areias$^{(\boxtimes)}$ and Ricardo Rocha

Faculty of Sciences, CRACS & INESC TEC, University of Porto,
Rua do Campo Alegre, 1021, 4169-007 Porto, Portugal
{miguel-areias,ricroc}@dcc.fc.up.pt

Abstract. Tabling is a technique that overcomes some limitations of traditional Prolog systems in dealing with redundant sub-computations and recursion. When tabling is combined with multithreading, we have the best of both worlds, since we can exploit the combination of higher declarative semantics with higher procedural control. To support this combination, the Yap Prolog system has, at engine level, multiple designs that vary from a No-Sharing design, where each thread allocates fully private tables, to a Full-Sharing (FS) design, where threads share the complete table space. In this work, we propose an extension to the table space data structures, which we named *Private Answer Chaining (PAC)*, as way to support batched scheduling evaluation with the FS design. Batched scheduling is one of the most successful tabling scheduling strategies, known to be useful when a tabled logic program requires an eager propagation of answers and/or do not requires the complete set of answers to be found. Experimental results show that PAC is a good first approach, since with it the FS design remains quite competitive.

Keywords: Logic programming · Multithreading · Tabling · Scheduling

1 Introduction

Tabling [5] is a technique that overcomes some limitations of traditional Prolog systems in dealing with redundant sub-computations and recursion. Tabling consists in storing intermediate answers for subgoals in a proper data structure, called the *table space*, so that they can be reused when a repeated subgoal appears during the resolution process. Tabling has become a popular and successful technique thanks to the ground-breaking work in the XSB Prolog system and in particular in the SLG-WAM engine [10], the most successful engine of XSB. Implementations of tabling are now widely available in systems like Yap Prolog, B-Prolog, ALS-Prolog, Mercury, Ciao Prolog and more recently Picat.

Multithreading in Prolog is the ability to concurrently perform computations, in which each computation runs independently but shares the program clauses. When multithreading is combined with tabling, we have the best of both worlds, since we can exploit the combination of higher procedural control with higher declarative semantics. To the best of our knowledge, XSB [8]

© Springer International Publishing Switzerland 2015
J.-L. Sierra-Rodríguez et al. (Eds.): SLATE 2015, CCIS 563, pp. 113–124, 2015.
DOI: 10.1007/978-3-319-27653-3_11

and Yap [2] are the only Prolog systems that support the combination of multithreading with tabling. In this work, we will focus on Yap's implementation, which follows a SWI-Prolog compatible multithreading library [11]. For tabled evaluation, a thread views its tables as private but, at the engine level, Yap has three designs [2], which vary from a *No-Sharing* (NS) design, where each thread allocates private tables for each new tabled subgoal call, to a *Full-Sharing* (FS) design, where threads share the complete table space.

The decision about the evaluation flow is determined by the *scheduling strategy*. Different strategies may have a significant impact on performance, and may lead to a different ordering of solutions to the query goal. Arguably, the two most successful tabling scheduling strategies are *local scheduling* and *batched scheduling* [6]. Local scheduling tries to complete subgoals as soon as possible. When new answers are found, they are added to the table space and the evaluation fails. Local scheduling has the advantage of minimizing the size of *clusters of dependent subgoals*, however it delays propagation of answers and requires the complete evaluation of the search space.

Batched scheduling favors forward execution first, backtracking next, and consuming answers or completion last. It thus tries to delay the need to move around the search tree by batching the return of answers to repeated subgoals. When new answers are found for a particular tabled subgoal, they are added to the table space and the evaluation continues. Batched scheduling can be an useful strategy in tabled logic programs that require an eager propagation of answers and/or do not require the complete set of answers to be found.

With the FS design, all tables are shared. Thus, since several threads can be inserting answers in the same table, when an answer already exists, it is not possible to determine if the answer is new or repeated for a particular thread without further support. For local scheduling, this is not a problem since, for repeated and new answers, local scheduling always fails. The problem is with batched scheduling that requires that only the repeated answers should fail. Threads have then to detect, during batched evaluation, whether an answer is new and must be propagated or whether an answer is repeated and the evaluation should fail.

In this work, we propose an extension to the table space data structures, which we named *Private Answer Chaining (PAC)*, as a way to keep track, per thread and subgoal call, of the answers that were already found and propagated. We discuss in detail our proposal for extending the FS design with batched scheduling and we present a performance analysis comparison between local and batched scheduling. Experimental results show that, despite the extra PAC data structures required to support batched scheduling with the FS design, the execution time of the combination is still quite competitive.

The remainder of the paper is organized as follows. First, we briefly introduce some background and related work. Then, we describe our PAC approach and we discuss the most important implementation details. Finally, we present experimental results and we end by outlining some conclusions.

2   Background

The basic idea behind tabling is straightforward: programs are evaluated by storing answers for tabled subgoals in an appropriate data space, called the *table space*. Repeated calls[1] to tabled subgoals are not re-evaluated against the program clauses, instead they are resolved by consuming the answers already stored in their table entries. During this process, as further new answers are found, they are stored in their tables and later returned to all repeated calls.

Figure 1 shows Yap's table space organization. At the entry point we have the *table entry* data structure. This structure is allocated when a tabled predicate is being compiled, so that a pointer to the table entry can be included in its compiled code. This guarantees that further calls to the predicate will access the table space starting from the same point. Below the table entry, we have the *subgoal trie structure*. Each different tabled subgoal call to the predicate at hand corresponds to a unique path through the subgoal trie structure, always starting from the table entry, passing by several subgoal trie data units, the *subgoal trie nodes*, and reaching a leaf data structure, the *subgoal frame*. The

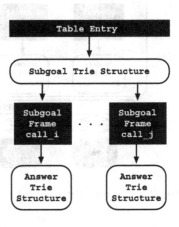

Fig. 1. Table space organization

subgoal frame stores additional information about the subgoal and acts like an entry point to the *answer trie structure*. Each unique path through the answer trie data units, the *answer trie nodes*, corresponds to a different answer to the entry subgoal.

2.1   Yap's Multithreaded Tabling Support

In Yap, a thread views its tables as private but, at the engine level, it implements three designs for concurrent tabling support that vary from a *No-Sharing* (NS) design, where each thread allocates fully private tables, to a *Full-Sharing* (FS) design, where threads share the complete table space. Figure 2 shows Yap's multithreaded table space organization for the NS and FS designs, where an interface layer abstracts the design being used at the engine level. The figure illustrates the main differences between the two designs for a situation where several threads are evaluating the same tabled subgoal call *call_i*.

When using the NS design, one can observe that the table entry data structure still stores the common information for the predicate (such as the arity or the scheduling strategy), and then each thread t has its own cell T_t inside a *bucket array* which points to the private data structures.

---

[1] We are considering *variant-based tabling* [9]. Two tabled subgoals A and B are variants if they can be made identical by variable renaming. For example, *p(X,1,Y)* and *p(Y,1,Z)* are variants because both can be transformed into $p(VAR_0, 1, VAR_1)$.

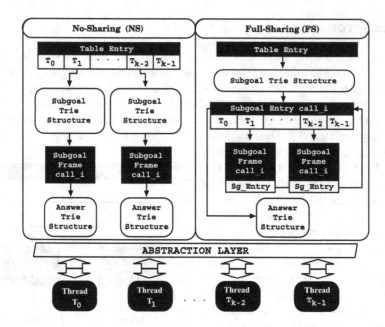

Fig. 2. Yap's multithreaded table space organization for the NS and FS designs

When using the FS design, the subgoal and answer trie structures and part of the subgoal frame (the *subgoal entry* data structure in Fig. 2) are shared among all threads. The previous subgoal frame data structure was split in two: the *subgoal entry* stores common information for the subgoal call (such as the pointer to the shared answer trie structure); the remaining information is kept private to each thread in the *subgoal frame* data structure. To support concurrency within the subgoal/answer tries, the FS design supports lock-based and lock-free approaches. A comparison between both approaches can be found in [3].

2.2   Scheduling Strategies

Local scheduling evaluates a tabled logic program in a breath-first manner. It favors backtracking first with completion instead of forward execution, leaving the consumption of answers for last. Local scheduling only allows a *Cluster of Dependent Subgoals (CDS)* to return answers after a fix-point has been reached [6]. In other words, local scheduling tries to keep a CDS as minimal as possible, thus creating less complex dependencies between subgoals, which causes a sooner completion of subgoals.

On the other hand, batched scheduling evaluates a tabled logic program in a depth-first manner. It favors forward execution first instead of backtracking, leaving the consumption of answers and completion for last. It thus tries to delay the need to move around the search tree by batching the return of answers. When new answers are found for a particular tabled subgoal, they are added to the table space and the execution continues. For some situations, this results in creating dependencies to older subgoals, therefore enlarging the current CDS and delaying the fix-point that guarantees that all dependent subgoals in a CDS are completely evaluated [10]. Batched scheduling can be an useful strategy in tabled logic programs that require an eager propagation of answers and/or do not require the complete set of answers to be found.

3 Extending Full-Sharing with Batched Scheduling

In this section, we describe our proposal to support the combination of batched scheduling with the FS design. In the original FS design, answer propagation and answer representation are both stored in the answer trie data structure, thus threads are unable to distinguish whether they have or not have propagated an answer already stored in the table space. To solve that, we propose an extension to the table space data structures, which we named *Private Answer Chaining* (PAC), as a way to keep track, per thread and subgoal call, of the answers that were already found and propagated to the thread's repeated calls. Figure 3 illustrates PAC's key idea. In a nutshell, PAC splits answer propagation from

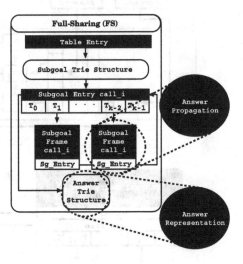

Fig. 3. PAC overview

answer representation, and allows the first to be privately stored in the subgoal frame data structure of each thread, and the second to be kept publicly shared among threads in the answer trie data structure.

3.1 Our Approach

The PAC procedure works at the subgoal frame level. The key idea is to extend subgoal frames with an auxiliary private chaining of answers for each subgoal call, in order to keep track of the answers already found for the call. Later, if a thread completes a subgoal's evaluation, i.e., if the subgoal's table is marked as complete, its PAC is made public, so that from that point on all threads can use that chain in complete (only reading) mode. Figure 4 illustrates the new data structures involved in the implementation of our PAC's proposal for a situation where different threads are evaluating the same tabled subgoal call *call_i*.

Figure 4(a) shows then a situation where two threads, T_1 and T_{k-2}, are sharing the same subgoal entry for a call *call_i* still under evaluation, i.e., still not yet completed. The current state of the evaluation shows an answer trie with

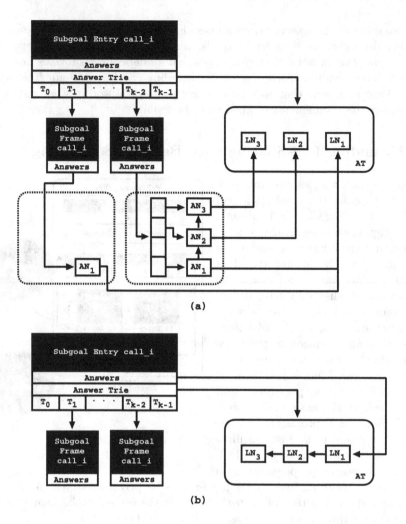

Fig. 4. PAC's data structures for (a) private and (b) public chaining

3 answers found for *call_i*. For the sake of simplicity, we are omitting the internal answer trie nodes and we are only showing the leaf nodes LN_1, LN_2 and LN_3 of each answer.

With PAC support, the leaf nodes are not chained in the answer trie data structure, as usual. Now, the chaining process is done privately, and for that, we use the subgoal frame structure of each thread. On the subgoal frame structure we added a new field, called *Answers*, to store the answers found within the execution of the thread. In order to minimize PAC's impact, each answer node in the private chaining has only two fields: (i) an entry pointer, which points to the corresponding leaf node in the answer trie data structure; and (ii) a next pointer to chain the nodes in the private chaining. To maintain good performance, when the number of answer nodes exceeds a certain threshold, we use a hash trie mechanism design similar to the one presented in [4], but without concurrency support, since this mechanism is private to each thread.

PAC's data structures in Fig. 4(a) represent then two different situations. Thread T_1 has only found one answer and it is using a direct answer chaining to access the leaf node LN_1. Thread T_{k-2} was already found three answers for $call\_i$ and it is using the hash trie mechanism within its private chaining. In the hash trie mechanism, the answer nodes are still chained between themselves, thus that repeated calls belonging to thread T_{k-2} can consume the answers as in the original mechanism.

Figure 4(b) shows the state of the subgoal call after completion. When a thread T completes a subgoal call, it frees its private consumer structures, but before doing that, it checks whether another thread as already marked the subgoal as completed. If no other thread has done that, then thread T not only follows its private chaining mechanism, as it would for freeing its private nodes, but also follows the pointers to the answer trie leaf nodes in order to create a chain inside the answer trie. Since this procedure is done inside a critical region, no more than one thread can be doing this chaining process. Thus, in Fig. 4(b), we are showing the situation where the subgoal call $call\_i$ is completed and both threads T_1 and T_{k-2} have already chained the leaf nodes inside the answer trie and removed their private chaining structures.

3.2   Implementations Details

The major difference between local and batched scheduling, at the engine level, is in the *tabled new answer* operation, where we decide what to do when a new answer is found during the evaluation. This operation checks whether a newly found answer is already in the corresponding answer trie structure and, if not, inserts it. For local scheduling, it then fails and, for batched scheduling, it proceeds with forward execution. Algorithm 1 shows how we have extended this operation to support the FS design with batched scheduling.

The algorithm receives two arguments: the newly found answer during the evaluation (ANS) and the subgoal frame which corresponds to the call at hand (SF). The algorithm begins by checking/inserting the given ANS into the answer trie structure, which will return the leaf node for the path representing ANS (line 1). Then, it checks/inserts the given *leaf* node into the private chaining for the current thread, which will return the corresponding answer chain node (line 2). Next in line 3, it tests whether the answer chain node already existed in the chain, i.e., if it was inserted or not by the current check/insert operation in order to return failure (line 4), or it proceeds with marking the answer ANS has found (line 6). At the end (lines 7 to 10), it returns failure, if local scheduling is active (line 8), otherwise, batched scheduling is active, and it proceeds by propagating the answer ANS to the current execution environment (line 10).

Algorithm 1. tabled_new_answer(answer ANS, subgoal frame SF)

1: $leaf \leftarrow check\_insert\_answer\_trie(ANS, SF)$
2: $chain \leftarrow check\_insert\_consumer\_chain(leaf, SF)$
3: **if** $is\_answer\_marked\_as\_found(chain)$ **then**
4:     **return** $failure$
5: **else** {the answer is new}
6:     $mark\_answer\_as\_found(chain)$
7:     **if** $local\_scheduling\_mode(SF)$ **then**
8:         **return** $failure$
9:     **else** {batched scheduling mode}
10:         **return** $proceed$

---

4   Experimental Results

We now present experimental results about the usage of PAC in the FS design with batched scheduling. The environment for our experiments was a machine with 32-Core AMD Opteron (TM) Processor 6274 (2 sockets with 16 cores each) with 32 GB of main memory, running the Linux kernel 3.16.7-200.fc20.x86_64 with Yap Prolog 6.3.

4.1   Benchmark Programs

For the experiments, we used the *TabMalloc* memory allocator [1] and five sets of benchmarks that create *worst case scenarios*, where we are able to show the lowest bounds of performance that each design might achieve when applied/used in other real world applications/programs. The *Large Joins* and *WordNet* sets were obtained from the OpenRuleBench project [7]; the *Model Checking* set includes three different specifications and transition relation graphs usually used in model checking applications; the *Path Left* and *Path Right* sets implement two recursive definitions of the well-known *path*/2 predicate, that computes the transitive closure in a graph, using several different configurations of *edge*/2 facts (Fig. 5 shows an example for each configuration). We experimented the *BTree* configuration with depth 17, the *Pyramid* and *Cycle* configurations with depth 2000 and the *Grid* configuration with depth 35.

In order to have a deeper insight on the behavior of each benchmark, and therefore clarify some of the results that are presented next, we first characterize the benchmarks. The columns in Table 1 have the following meaning:

- **calls:** is the number of different calls to tabled subgoals. It corresponds to the number of paths in the subgoal tries.
- **trie nodes:** is the total number of trie nodes allocated in the corresponding subgoal/answer trie structures.
- **trie depth:** is the minimum/average/maximum number of trie nodes required to represent a path in the corresponding subgoal/answer trie structures. Trie structures with smaller average depth values are more amenable to higher contention.

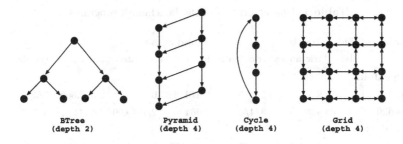

| BTree | Pyramid | Cycle | Grid |
| (depth 2) | (depth 4) | (depth 4) | (depth 4) |

Fig. 5. Edge configurations for the path benchmarks

- **unique:** is the number of different tabled answers found. It corresponds to the number of paths in the answer tries.
- **repeated:** is the number of redundant tabled answers found.

By observing Table 1, the *Mondial* benchmark, from the *Large Joins* set, and the three *Model Checking* benchmarks seem to be the benchmarks least amenable to contention since they are the ones that find less unique answers and that have the deepest trie structures. In this regard, the *Path Left* and *Path Right* sets correspond to the opposite case. They find a huge number of answers and have very shallow trie structures. On the other hand, the *WordNet* and *Path Right* sets have the benchmarks with the largest number of different subgoal calls, which can reduce the probability of contention because answers can be found for different subgoal calls and therefore be inserted with minimum overlap. On the opposite side are the *Join2* benchmark, from the *Large Joins* set, and the *Path Left* benchmarks, which have only a single tabled subgoal call.

4.2 Performance Analysis

We present now the performance analysis about the usage of PAC in the FS design with batched scheduling. To support concurrency within the subgoal/answer tries, the FS design is using the lock-free hash trie design presented in [3]. Since without PAC the FS design would not be able to be used with batched scheduling, to put PAC's results in perspective, we will be showing also the results for local scheduling and for the NS design.

Table 2 shows the overhead ratios for the five sets of benchmarks, when comparing against the NS design with 1 thread (running with local scheduling and without TabMalloc), for the NS and FS designs with 1, 8, 16, 24 and 32 threads, using local scheduling (column *Local*) and batched scheduling (column *Batched*) strategies with TabMalloc. In order to give a fair weight to each benchmark, the overhead ratio is calculated as follows. We begin by running 10 times each benchmark B for each design D with T threads. Then, we calculate the average of those ten runs and use that value (D_{BT}) to put it in perspective against the base time, which is the average of 10 runs of the NS design with 1 thread (NS_{B1}). For that, we use the following formula for the overhead $O_{DBT} = D_{BT}/NS_{B1}$.

Table 1. Characteristics of the benchmark programs

Bench	Tabled subgoals			Tabled answers			
	calls	trie nodes	trie depth	unique	repeated	trie nodes	trie depth
Large Joins							
Join2	1	6	5/5/5	2,476,099	0	2,613,660	5/5/5
Mondial	35	42	3/4/4	2,664	2,452,890	14,334	6/7/7
WordNet							
Clusters	117,659	235,319	2/2/2	166,877	161,853	284,536	1/1/1
Hypo	117,657	117,659	2/2/2	698,472	20,341	816,129	1/1/1
Holo	117,657	235,315	2/2/2	74,838	54	192,495	1/1/1
Hyper	117,657	235,315	2/2/2	698,472	8,658	816,129	1/1/1
Tropo	117,657	235,315	2/2/2	472	0	118,129	1/1/1
Mero	117,657	117,659	2/2/2	74,838	13	192,495	1/1/1
Model Checking							
IProto	1	6	5/5/5	134,361	385,423	1,554,896	4/51/67
Leader	1	5	4/4/4	1,728	574,786	41,788	15/80/97
Sieve	1	7	6/6/6	380	1,386,181	8,624	21/53/58
Path Left							
BTree	1	3	2/2/2	1,966,082	0	2,031,618	2/2/2
Pyramid	1	3	2/2/2	3,374,250	1,124,250	3,377,250	2/2/2
Cycle	1	3	2/2/2	4,000,000	2,000	4,002,001	2/2/2
Grid	1	3	2/2/2	1,500,625	4,335,135	1,501,851	2/2/2
Path Right							
BTree	131,071	262,143	2/2/2	3,801,094	0	3,997,700	1/2/2
Pyramid	3,000	6,001	2/2/2	6,745,501	2,247,001	6,751,500	1/2/2
Cycle	2,001	4,003	2/2/2	8,000,000	4,000	8,004,001	1/2/2
Grid	1,226	2,453	2/2/2	3,001,250	8,670,270	3,003,701	1/2/2

After calculating all the overheads O_{DBT} for a certain design D and number of threads T corresponding to the several benchmarks B, we calculate the respective minimum, average, maximum and standard deviation overhead ratios (rows *Min*, *Avg*, *Max* and *StD* in Table 2).

By observing Table 2, we can see that batched scheduling always achieves the best minimum overhead ratio in the FS design but, for the average and maximum overhead ratios, the best strategy is always local scheduling. For the average and maximum overhead ratios, the difference between local and batched scheduling in the FS design is slightly higher than in the NS design, which can be read as an indication of the overhead that PAC introduces into the FS design. Recall that whenever an answer is found during the evaluation, PAC requires that threads traverse their private consumer data structures to check if the answer was already found (and propagated).

Table 2. Overhead ratios, when compared with the NS design with 1 thread (running with local scheduling and without TabMalloc) for the NS and FS designs (with Tab-Malloc) when running 1, 8, 16, 24 and 32 threads with local and batched scheduling (best ratios by row and design for the Minimum, Average and Maximum are in bold)

Threads		NS		FS	
		Local	Batched	Local	Batched
1	Min	**0.53**	0.55	1.01	**0.95**
	Avg	**0.78**	0.82	**1.30**	1.46
	Max	1.06	**1.05**	**1.76**	2.33
	StD	0.15	0.14	0.22	0.44
8	Min	0.66	**0.63**	1.16	**0.99**
	Avg	**0.85**	0.88	**1.88**	1.95
	Max	**1.12**	1.14	**2.82**	3.49
	StD	0.13	0.14	0.60	0.79
16	Min	0.85	**0.75**	1.17	**1.06**
	Avg	**0.98**	1.00	**1.97**	2.08
	Max	**1.16**	1.31	**3.14**	3.69
	StD	0.09	0.17	0.65	0.83
24	Min	**0.91**	0.93	1.16	**1.09**
	Avg	**1.15**	1.16	**2.06**	2.19
	Max	1.72	**1.60**	**3.49**	4.08
	StD	0.20	0.21	0.70	0.91
32	Min	1.05	**1.04**	1.33	**1.26**
	Avg	1.51	**1.49**	**2.24**	2.41
	Max	**2.52**	2.63	**3.71**	4.51
	StD	0.45	0.45	0.74	1.02

As we increase the number of threads, for the NS design, both scheduling strategies show very close minimum, average and maximum overhead ratios. For the FS design, the differences are slightly higher. However, for the average overhead ratio, the results between both strategies are quite close, with batched scheduling being around 10 % slower than local scheduling for the FS design. In summary, our experimental results show that, on average, the PAC strategy does not seem to have a big impact in the performance, however it still leaves room for further improvements, since the difference between local and batched scheduling is higher in the FS design than in the NS design.

5   Conclusions and Further Work

Local and batched scheduling are arguably two of the most well-known tabling scheduling strategies. The major difference between both is that local scheduling propagates answers only after all answers are found, while batched scheduling propagates answers immediately after they are found. Batched scheduling is a useful strategy in tabled logic programs that require an eager propagation of answers and/or do not require the complete set of answers to be found. In this work, we have presented the PAC strategy, which is a simple and novel approach for combining the FS design with batched scheduling. PAC splits answer representation from answer propagation, and allows the first to be publicly shared among threads while the second to be private to each thread.

Experimental results in worst-case scenarios showed that, on average, the PAC strategy does not seem to have a big impact in the performance, however it still leaves room for further improvements specially in the extra structures required to control the propagated answers. Further work will include the usage of time-stamped tries to minimize the search for the propagated answers and new real-world problems that will allow us to improve and consolidate our framework.

Acknowledgments. This work is partially funded by the North Portugal Regional Operational Programme (ON.2 - O Novo Norte) and by the National Strategic Reference Framework (NSRF), through the European Regional Development Fund (ERDF) and the Portuguese Foundation for Science and Technology (FCT), within projects NORTE-07-0124-FEDER-000059 and UID/EEA/50014/2013.

References

1. Areias, M., Rocha, R.: An efficient and scalable memory allocator for multithreaded tabled evaluation of logic programs. In: International Conference on Parallel and Distributed Systems, pp. 636–643. IEEE Computer Society (2012)
2. Areias, M., Rocha, R.: Towards multi-threaded local tabling using a common table space. J. Theory Pract. Logic Program. **12**(4 & 5), 427–443 (2012)
3. Areias, M., Rocha, R.: A simple and efficient lock-free hash trie design for concurrent tabling. In: Technical Communications of the International Conference on Logic Programming (2014)
4. Areias, M., Rocha, R.: A lock-free hash trie design for concurrent tabled logic programs. Int. J. Parallel Program. 1–21 (2015)
5. Chen, W., Warren, D.S.: Tabled evaluation with delaying for general logic programs. J. ACM **43**(1), 20–74 (1996)
6. Freire, J., Swift, T., Warren, D.S.: Beyond depth-first: improving tabled logic programs through alternative scheduling strategies. In: Kuchen, H., Swierstra, S.D. (eds.) PLILP 1996. LNCS, vol. 1140, pp. 243–258. Springer, Heidelberg (1996)
7. Liang, S., Fodor, P., Wan, H., Kifer, M.: OpenRuleBench: an analysis of the performance of rule engines. In: International World Wide Web Conference, pp. 601–610. ACM (2009)
8. Marques, R., Swift, T.: Concurrent and local evaluation of normal programs. In: Garcia de la Banda, M., Pontelli, E. (eds.) ICLP 2008. LNCS, vol. 5366, pp. 206–222. Springer, Heidelberg (2008)
9. Ramakrishnan, I.V., Rao, P., Sagonas, K., Swift, T., Warren, D.S.: Efficient access mechanisms for tabled logic programs. J. Logic Program. **38**(1), 31–54 (1999)
10. Sagonas, K., Swift, T.: An abstract machine for tabled execution of fixed-order stratified logic programs. ACM Trans. Program. Lang. Syst. **20**(3), 586–634 (1998)
11. Wielemaker, J.: Native preemptive threads in SWI-prolog. In: Palamidessi, C. (ed.) ICLP 2003. LNCS, vol. 2916, pp. 331–345. Springer, Heidelberg (2003)

Browsing the Parse Space

Daniel Rodríguez-Cerezo and José-Luis Sierra[✉]

Fac. Informática, Universidad Complutense de Madrid, Madrid, Spain
{drcerezo, jlsierra}@fdi.ucm.es

Abstract. Ambiguous context-free grammars can generate many (even infinite) parse trees for each input sentence. We will refer to all these parse trees as the *parse space* of the sentence. Thus, in many settings (computational linguistics, education in compiler construction, etc.) the need for *browsing* this parse space (i.e., for examining different trees in a systematic and ordered way) arises. In this paper we describe a browsing approach that works for arbitrary (even infinitely ambiguous) grammars. The approach, which is based on the well-known Earley's algorithm, sorts the parse space according to structural complexity of the parse trees, lets users inspect a particular tree, and then to jump to the previous and/or the next tree. This approach has been implemented in *EvDebugger*, an educational system for the learning of the attribute grammar formalism.

Keywords: Parsing · Earley algorithm · Parse trees · Grammar debugging technique

1 Introduction

Nowadays context-free grammars are keystone artifacts for specifying the syntax of both natural and artificial languages [11]. Using context-free grammars, language engineers can describe the structural concepts of a language, establishing in this way the basic skeleton in which to base all the subsequent processing activities. In addition, they can use standard grammar analysis algorithms to check desired properties of the proposed grammar (e.g., whether the grammar is *proper* –i.e., whether all the symbols are *accessible* from the initial symbol as well as *productive*) [1]. Unfortunately, due to their expressive power, many interesting properties become undecidable [4]. One of these properties is *ambiguity*, i.e., whether the grammar is able to impose several, more than one, alternative structures (i.e., *parse trees*) to a sentence.

Ambiguity is a typical phenomenon to avoid when modeling artificial languages. For this purpose, ambiguity sources must be clearly detected and fixed. These sources can be due to some formalization leak (in this case, they can be solved by changing the grammar), or they can be rooted in the primary conception of the language itself (in this case, it is the language conception that must be changed). A way of avoiding ambiguity is to circumscribe the class of allowable grammars to one for which unambiguity can be ensured (e.g., LL or LR grammars). However, the membership tests for these classes provide little information on the ambiguity sources (on the contrary, they identify sources of non-determinism in the associated parsing algorithms, which may or may not be related to ambiguity). Thus, when analyzing ambiguity language engineers can

© Springer International Publishing Switzerland 2015
J.-L. Sierra-Rodríguez et al. (Eds.): SLATE 2015, CCIS 563, pp. 125–136, 2015.
DOI: 10.1007/978-3-319-27653-3_12

take benefit of a more empirical approach, by selecting representative sentences and by examining the *parse spaces* of these sentences (i.e., the set of the sentences' parse trees –see Fig. 1, a grammar that corresponds to a real case concerning a typical misleading among compiler construction students at UCM: the aim was to model optional sequences of items, e.g., instructions; a common trend among students was to make the items optional instead the whole sequence). This approach can be particularly valuable for educational purposes by letting students of compiler construction courses *visualize* the ambiguity phenomenon and identify the potential sources of ambiguity (for instance, the trees in Fig. 1 make apparent how the source of ambiguity is the afore-mentioned misleading of making individual items optional instead the sequence itself). In addition, the ability of visualizing and inspecting ambiguity can be also very valuable in natural language settings, where ambiguity is not a phenomenon to avoid but an intrinsic feature of natural languages [9].

(a)   (b)

$L \rightarrow L\,I \mid I$
$I \rightarrow i \mid \lambda$

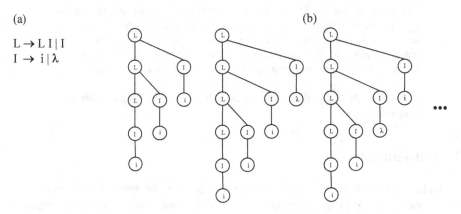

Fig. 1. (a) An ambiguous context-free grammar; (b) parse space associated with the sentence iii with respect to the grammar in (a)

In order to support the aforementioned empirical approach, in this paper we describe a strategy for systematically browsing the parse space of a sentence according to an arbitrary context-free grammar (even infinitely ambiguous: for instance, the grammar in Fig. 1(a) is infinitely ambiguous, since the parse space in Fig. 1(b) for the sentence iii contains infinitely many trees; dealing with this kind of grammars leaves out *naif* approaches, such as generating a list with all the possible parse trees as a previous step to browsing). The strategy uses the well-known Earley's algorithm [10] to recognize the input sentence, and then it exploits the Earley's list produced by the algorithm to lazily enumerate all the possible trees. In this enumeration, simpler trees are generated first. Generated trees are backed up for letting users move to previous trees. This strategy has been successfully implemented in *EvDebugger* [13], an edu-cational system focused on the attribute grammar formalism, in order to let students browse the parse space of the input sentence and examine the attribute evaluation process on each visited tree.

(a)

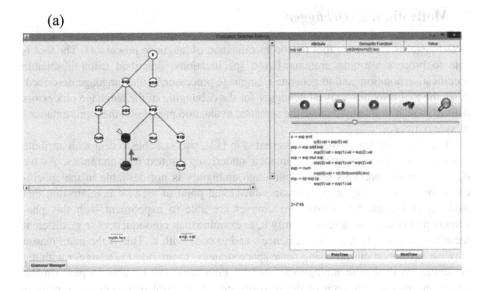

(b)

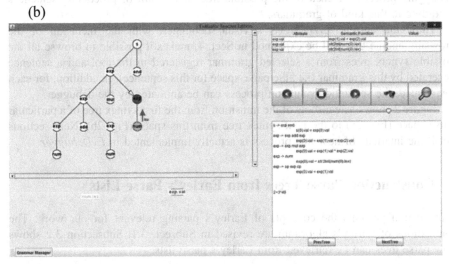

Fig. 2. Browsing the parse space with *EvDebugger*

The rest of the paper is organized as follows: Sect. 2 motivates the work by showing how *EvDebugger* deals with ambiguous grammars. Section 3 describes how parse trees can be constructed form Earley's parse lists. Section 4 describes the browsing engine. Section 5 outlines some related work. Finally, Sect. 6 presents some conclusions and outlines some lines of future work.

2   Motivation: *EvDebugger*

EvDebugger is a software tool for the specification of language processors. The tool is able to process attribute grammar-based specifications, described using a suitable specification notation, and to generate a language processor for the language described. Also, the tool provides a visual debugger for the debugging of the language processors defined, which is able to animate the semantic evaluation process on the representations of syntax trees.

The initial version of the tool, presented in [13], was not able to deal with attribute grammars that present ambiguity in their underlying context-free grammars. As we discussed in the previous section, although ambiguity is not desirable in the specification of artificial languages, from an educational point of view, it is interesting that students of Compiler Constructions courses are able to experiment with this phenomenon in order to be able to identify it, to examine their consequences (e.g., different meanings assigned to the same sentence), and to cope with it. This is the main reason why we developed our browsing parse space strategy, to provide *EvDebugger* with the possibility to deal with ambiguous attribute grammars and be able to perform the debugging process for each of the possible trees that result of processing sentences according to this kind of grammars.

Thus, the current *EvDebugger*'s visual debugger, with the inclusion of the browsing engine that will be explained in Sect. 4, makes it possible to browse all the possible syntax trees from a selected grammar registered in the tool and a sentence generated by this grammar (i.e., the parse space for this sentence). In addition, for each tree visited, the semantic evaluation process can be animated by the debugger.

Figure 2 shows screenshots of the transition from the first syntax tree of a particular parse space (Fig. 2a) to the next syntax tree from this space (Fig. 2b). Next sections detail the internals of how this process is actually implemented in *EvDebugger*.

3   Constructing Parse Trees from Earley's Parse Lists

This section presents the concepts of Earley's parsing relevant for our work. The foundations of Earley's algorithm are revised in Subsect. 3.1. Subsection 3.2 shows how parse trees can be retrieved from Earley's parse lists.

3.1   Earley's Recognizer

Earley's algorithm can be though as a way of computing, for an input sentence w, a set of items of the form $<i,j, A \rightarrow \alpha \bullet \beta >$, where i and j are natural numbers, and $A \rightarrow \alpha\beta$ is a syntax rule[1]. The intended meaning of these items is: (i) the input fragment $w[j \ldots i\text{-}1]$ can be derived from α, (ii) it may be followed by another fragment y that can be derived from β, and thus (iii) $w[j \ldots i\text{-}1]y$ may be derived from A.

---

[1] In the Earley's original work lookahead symbols were also added to items, although later on it was shown that it does not substantially affect to algorithm performance.

init: $<1,1,S' \to \bullet S >$

predictor: $\dfrac{<i,j, A \to \alpha \bullet B\ \beta> \;;;\ B \to \gamma}{<i,i, B \to \bullet \gamma>}$

scanner: $\dfrac{<i,j, A \to \alpha \bullet a\ \beta> \;;;\ w[i] = a}{<i+1, j, A \to \alpha\ a\bullet \beta>}$

completer: $\dfrac{<i, j, B \to \gamma \bullet > \;;;\ < j, k, A \to \alpha \bullet B\ \beta >}{<i, k, A \to \alpha\ B\bullet \beta>}$

Fig. 3. Earley's recognition calculus (by $<i,j,A\to\alpha\bullet\beta>$ we denote an item in the itemset, by $A\to\alpha$ a syntax rule, by S the grammar initial symbol, and by S' the initial symbol of the augmented grammar)

(1) $<1,1,L' \to \bullet L>^{I}$	(10) $<2,1,I \to i\bullet>^{s(4)}$
(2) $<1,1,L \to \bullet LI>^{p(1)}$	(11) $<2,1,L \to I\bullet>^{c(10,3)}$
(3) $<1,1,L \to \bullet I>^{p(1)}$	(12) $<2,1,L \to L\bullet I>^{c(10,2)\ c(15,2)}$
(4) $<1,1,I \to \bullet i>^{p(3)}$	(13) $<2,2,I \to \bullet i>^{p(12)}$
(5) $<1,1,I \to \bullet>^{p(3)}$	(14) $<2,2,I \to \bullet>^{p(12)}$
(6) $<1,1,L \to I\bullet>^{c(5,3)}$	(15) $<2,1,L \to LI\bullet>^{c(10,7)\ c(14,12)}$
(7) $<1,1,L \to L\bullet I>^{c(6,2)\ c(8,2)}$	(16) $<2,1,L' \to L\bullet>^{c(15,1)}$
(8) $<1,1,L \to LI\bullet>^{c(6,7)}$	
(9) $<1,1,L' \to L\bullet>^{c(8,1)}$	

(17) $<3,2,I \to i\bullet>^{s(13)}$	(23) $<4,3,I \to i\bullet>^{s(20)}$
(18) $<3,1,L \to LI\bullet>^{c(17,12)\ c(21,19)}$	(24) $<4,1,L \to LI\bullet>^{c(23,19)\ c(27,25)}$
(19) $<3,1,L \to L\bullet I>^{c(18,2)}$	(25) $<4,1,L \to L\bullet I>^{c(24,2)}$
(20) $<3,3,I \to \bullet i>^{p(19)}$	(26) $<4,4,I \to \bullet i>^{p(25)}$
(21) $<3,3,I \to \bullet>^{p(19)}$	(27) $<4,4,I \to \bullet>^{p(25)}$
(22) $<3,1,L' \to L\bullet>^{c(18,1)}$	(28) $<4,1,L' \to L\bullet>^{c(24,1)}$

Fig. 4. Earley's parse list for sentence iii and grammar in Fig. 1a. Each item has a unique number assigned, as well as the set of rules generating it (*i* stands for the *init* rule, *p* for *predict* rule, *c* for *completer* and *s* for *scanner*; the items involved in the application of each rule are indicated as arguments –e.g, $c(23,19)$ stands for the application of the *completer* rule on items 23 and 19)

The calculus in Fig. 3 characterizes the Earley's item set associated with a sentence w with respect to a grammar. Earley's algorithm implements this recognition calculus by grouping items by their first component, and by disposing these groups in a *parse list*. There will be $|w|+1$ groups, in such as way w is accepted by the algorithm when the group $|w|+1$ contains an item of the form $<1,|w|+1, S' \to S \bullet>$ ($S'\to S$, where S is the grammar's initial symbol and S' is a new fresh initial symbol, is a convenience rule added to facilitate acceptance recognition as well as the collection of parse trees). In addition, when constructing the parse list, it is possible to add, to each item, information on the rules in Fig. 3 that generated it (in case of ambiguous grammars, an item could be generated by more than a rule in the recognition calculus). This information will be subsequently useful for recovering parse trees. As an example, Fig. 4 shows the Earley's parse list corresponding to the grammar and the sentence in Fig. 1.

3.2   The Tree Construction Calculus

Earley's parse list contains enough information for recovering any parse tree for the input sentence. Indeed, the recovering process can be characterized by a *tree construction calculus* like the shown in Fig. 5. This calculus models how to associate subtree sequences with Earley's items. For this purpose, it comprises judgements of the form i⊢τ, with *i* an Earley's item and τ a sequence of parse subtrees. The calculus itself can be derived in a straightforward way by examining the *scanner* and *completer* rules in the recognition calculus.

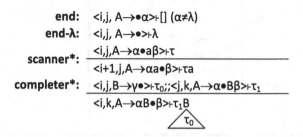

Fig. 5. Tree construction calculus

The tree construction calculus plays a primary role in our browsing approach. Indeed, proof trees in this calculus (*tree construction proof trees*) show how parse trees can be effectively constructed. Figure 6 shows an example of tree construction proof tree for the `iii` sentence and the grammar in Fig. 1.

Notice that tree construction proof trees can be built in two steps:

- First rules are applied in a top-down way to determine the judgement antecedents in the proof tree (i.e., the *i* part in a judgment i⊢τ). We will call to such a skeleton of tree construction proof tree a *tree construction plan*. Notice that by adding information about the application of the *completer* and *scanner* rules to the Earley's items (see Fig. 4), the Earley's parse lists will encode all the possible tree construction plans in a compact way.
- Then, judgment consequents (i.e, the τ part) are synthetized in a bottom-up way.

4   The Browsing Engine

The browsing engine lets users explore the parse spaces for arbitrary context-free grammars and input sentences. For this purpose, the engine organizes the parse spaces as a lazily-generated double-linked list of parse trees ordered by number of nodes.

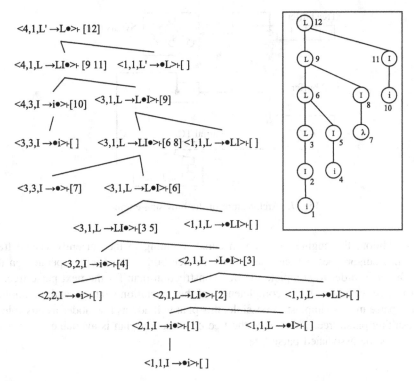

Fig. 6. Example of a tree construction proof tree (rule names are omitted by the sake of simplicity; instead they can be inferred from the number of child nodes. In addition, sequences of subtrees are represented as sequences of node numbers in the final parse tree –this parse tree appears inside the box).

For this purpose, and based on the tree construction calculus described in the previous section, the engine maintains a set of partial tree construction plans. When it completes one of these plans, it uses such a plan to synthetize the corresponding parse tree. Next subsections describe the details of this engine. The architecture and the browsing strategy is described in Subsect. 4.1. Subsection 4.2 describes how the engine completes the next tree construction plan. Finally, Subsect. 4.3 described how it synthetizes parse trees from tree construction plans.

4.1 Browser Architecture and Browsing Strategy

The browsing engine maintains two different buffers (see Fig. 7):

- A *Tree Buffer*. This buffer keeps a double-linked list with the parse trees already generated by the browsing process.
- A *Tree Construction Plan* (TCP) *Buffer*. This buffer indexes the partially-generated tree construction plans. This is initialized with the index elements corresponding to the item $\langle 1,|w|+1,S'\rightarrow S\bullet\rangle$.

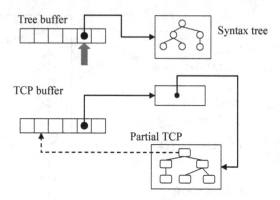

Fig. 7. Architecture of the browsing engine.

In addition, the engine maintains a cursor pointing to the currently visited tree. When this cursor goes beyond the end of the tree buffer, the engine operates on the TCP Buffer in order to complete the tree construction plan for the next parse tree. In order to ensure, on one hand, completeness of the generation strategy, and, on another hand, a parse tree as simplest as possible, this plan will add as few nodes as possible to the resulting parse tree. Thus, once the tree construction plan is available, the engine synthetizes the associated parse tree.

4.2   Getting the Next Tree Construction Plan

The generation of the next tree construction plan is performed using a breath-first generation strategy. For this purpose, the TCP buffer contains index elements of the form $<n, Os, Ls>$, where:

- n is the number of nodes added by the partial tree construction plan to the final parse tree.
- Os is the set of yet unexpanded Earley items in this plan.
- Ls is the set of nodes generated by the *end* or by the *end-λ* rule.

These elements are ordered by the number of nodes in increasing order. Thus, in order to get the next tree construction plan:

- The engine extracts the first element $<n, Os, Ls>$ of the TCP buffer.
- If Os is empty, it stops (the extracted element indexes the desired tree construction plan).
- Otherwise, it extract one item from Os, applies all the possible tree construction rules on it (this step is speeded-up by the information on the generating rules stored with the Earley's items), construct new indexing elements with the result, and inserts these elements in the TCP buffer (these elements are inserted in the appropriate positions to maintain the elements in this buffer ordered by the number of nodes).

```
nextTCP:

found:=false;
do
  <n,Os,Ls> := pop(TCP);
  if
  ¬ found →
    if
    Os=∅ → found:=true; []
    Os≠∅ →
      θ^Rs := pop(Os);
      if
      Rs = ⊥→
        if
        θ=<_,_,A→•α> →
          newNode = new Init(parent:θ^Rs);
          insert(TCP,<n+1,Os, Ls++newNode>); []
        θ=<_,_, A→•> →
          newNode = new Init-λ(parent:θ^Rs);
          insert(TCP,<n+1,Os, Ls++newNode>);
        end if []
      Rs ≠ ⊥ →
```

```
        foreach R in Rs do
          if
          R= scanner(σ) →
            newRuleNode := new Scanner*(parent:θ^Rs);
            newItemNode := new Item(item: σ,
                                parent:newRuleNode);
            insert(TCP, <n+1,
                        Os++newItemNode, Ls>); []
          R= completer(σ,ω) →
            newRuleNode :=
                  new Completer*(parent:θ^Rs);
            newItemNode 1 := new Item(item:σ,
                            parent:newRuleNode);
            newItemNode2 := new Item(item:ω,
                            parent:newRuleNode);
            insert(TCP, <n+1,Os++[newItemNode1,
                            newItemNode2], Ls>);
          end if
        end foreach
        end if
      end if []
      found → return <n,Os,Ls>
    end if
  end do
```

Fig. 8. Pseudo-code for getting the next complete parse tree construction plan.

This behavior is detailed by the pseudo-code of Fig. 8. Notice that partial tree construction plans are represented with the arcs reversed (i.e., from children to parents).

4.3   Synthetizing the Parse Tree

As indicated in Sect. 4.1, once the tree construction plan is available, it is possible to synthetize the corresponding parse tree. Since the browsing engine represents plans with reversed arcs, it can be meaningfully addressed as a data-driven attribute evaluation process (see, for instance, [12]). For this purpose:

- Each *completer** rule node has a pointer to its first visited antecedent.
- To compute the subtree sequence of the parent on one of these nodes (the rule consequent), this pointer must be set when the node is visited.

The process itself is initiated by the *Ls* elements in the plan index $<n,∅,Ls>$ (the leaves of the plan) returned by the procedure described in the previous subsection, and it ends when the plan's root is reached. Since plans can share structure, it also takes care of restarting the *completer** rule *first visited antecedent* pointers when those nodes are fired. The strategy itself is formalized by the pseudo-code of Fig. 9.

5   Related Work

The typical way of dealing with parse spaces is by constructing *parse forests* (i.e., compact representations of all the possible parse trees of a sentence) [7]. The original work of Earley [10] suggested a way of building parse forests from the references

Tree synthesis for <$n,\varnothing,Ls$>:

```
PendingNodes = [];
foreach R in Ls
  if
    R = init(θ) →
      θ.τ := [];
      PendingNodes := PendingNodes++[θ]; []
    Rule = init-λ(θ) →
      θ.τ := [λ];
      PendingNodes := PendingNodes++[θ];
  end if
end foreach;
Node := pop(PendingNodes);
do
  Node.parent = ⊥→return Node.τ; []
  Node.parent ≠ ⊥→
    if
      Node.parent = scanner*(θ) →
        let Node = <_,_,A→αa•β> in
          θ.τ := Node.τ ++ a
        end let ;
        PendingNodes := PendingNodes++[θ]; []
```

```
      Node.parent = completer*(θ) →
        if
          Node.parent.theOtherArg=⊥→
            Node.parent.theOtherArg=Node; []
          Node.parent.theOtherArg≠⊥→
            if
              Node.parent.theOtherArg = <_,_,B→γγ•>→
                let Node = <_,_,A→α•Bβ> in
                  θ.τ := Node.τ ++ B
                    Node.parent.theOtherArg.τ ; []
              end let
              Node.parent.theOtherArg=<_,_,A→α•Bβ> →
                let Node = <_,_,B→γγ•> in
                  θ.τ := Node.τ ++ B
                    Node.parent.theOtherArg.τ ;
              end let
            end if
            Node.parent.theOtherArg := ⊥;
            PendingNodes := PendingNodes++[θ];
        end if
      end if
      Node := pop(PendingNodes);
end do
```

Fig. 9. Pseudo-code for the synthesis of parse trees from tree construction plans.

associated to the items in the parse lists. However, Tomita in [17] notices that Earley's method could be incorrect. The Tomita's parse method, as well as their successors in the GLR parse branch, incorporates parse forest construction as an essential feature. The works in [14, 15] shows how to adapt the Earley's parsing style to produce parse forests of the input sentences in a correct way. On the contrary to these approaches, ours does not attempt to build compact representations of parse spaces, but to browse these spaces. The browsing strategy presented in this paper could be adapted to work on parse forests instead of on parse lists, however, without substantial modifications. Nevertheless, it could introduce an unnecessary intermediate step, so we find our solution, based on the tree construction calculus, more natural and straightforward.

One of the potential applications of our approach is as a tool for helping language engineers to understand grammar ambiguity. Regardless its undecidable nature, ambiguity analysis is amenable of being addressed by approximation strategies. In [6] parse forests are analyzed for detecting typical causes of ambiguity. In [18] a non-exhaustive breath-first search strategy on the sentences generated by the grammar is used. The work in [8] uses regular approximations of the grammars under study to turn the problem of ambiguity detection in a decidable one. In [5] a usability analysis of the techniques involved in ambiguity detection is carried out. While all these techniques are oriented to automatize ambiguity detection and/or diagnose, our approach is more agnostic and user-centered, providing users with a tool that can be used for browsing the parse space, and which can be useful to diagnose the possible causes of ambiguity of a given grammar construction.

Finally, our approach can be useful in educational settings, in order to help students to better appreciate and understand the ambiguity phenomenon. For this purpose, it differs from other visualization-based comprehension approaches to parsing (e.g., [2, 3]), since these approaches are usually focused to visualize/animate the construction of single parse trees, instead of being focused on whole parse spaces. A preliminary version of our approach was implemented in PAG (*Prototyping with Attribute Grammars*) [16]. However, this preliminary implementation failed to work with infinitely ambiguous grammars. As aforementioned, the approach in its current form has been recently implemented in the *EvDebugger* system [13].

6 Conclusions and Future Work

In this paper, we have presented a strategy that allows the navigation of the parse space imposed by a grammar on a sentence. The strategy is built on the well-known Earley's algorithm and works for arbitrary (even infinitely ambiguous) context-free grammars. The approach sorts parse tree by structural complexity (first the simpler ones), and it is based on a breath-first construction of the proof trees for a parse tree construction calculus derived from the Earley's parsing calculus. This strategy has been implemented in *EvDebugger*, an educational IDE for language processor generation based on attribute grammars enriched with a visual debugger, in order to provide the capabilities needed to deal with the ambiguity of underlying context-free grammars.

Currently we are planning to carry out an empirical evaluation of the approach with students of a compiler construction course, in order to assess its usability as well as its efficacy as a tool for comprehending the ambiguity phenomenon and for diagnosing causes of ambiguity. We are also planning to carry out an in-depth analysis of efficiency criteria concerning the browsing approach and also to explore alternative construction strategies (e.g., using iterative deepening in connection with the tree construction calculus).

Acknowledgements. This work has been partially supported by the BBVA Foundation (research grant HUM14_251), by the Spanish R&D&I Plan (research grant TIN2014-52010-R), by Santander-UCM GR3/14 (group number 962022) and by the grant EDU/3445/201.

References

1. Aho, A.V., Ullman, J.D.: The Theory of Parsing, Translation, and Compiling, vol I: Parsing. Prentice-Hall, Inc, Englewood Cliffs (1972)
2. Almeida-Martınez, F.J., Urquiza-Fuentes, J., Velázquez-Iturbide, J.: Visualization of syntax trees for language processing courses. J. Univ. Comput. Sci. **15**(7), 1546–1561 (2009)
3. Almeida-Martínez, F.J., Urquiza-Fuentes, J., Velázquez-Iturbide, J.Á. VAST: visualization of abstract syntax trees within language processors courses. In: Proceedings of the 4th ACM Symposium on Software Visualization, pp. 209–210. ACM. (2008)
4. Bar-Hillel, Y., Perles, M., Shamir, E.: On formal properties of simple phrase structure grammars. STUF-Lang. Typology Univ. **14**(1–4), 143–172 (1961)
5. Basten, H.J.: The usability of ambiguity detection methods for context-free grammars. Electron. Notes Theor. Comput. Sci. **238**(5), 35–46 (2009)

6. Basten, H.J., Vinju, J.J.: Parse forest diagnostics with Dr. Ambiguity. In: Sloane, A., Aßmann, U. (eds.) SLE 2011. LNCS, vol. 6940, pp. 283–302. Springer, Heidelberg (2012)
7. Billot, S., Lang, B.: The structure of shared forests in ambiguous parsing. In: Proceedings of the 27th Annual Meeting on Association for Computational Linguistics, pp. 143–151. Association for Computational Linguistics (1989)
8. Brabrand, C., Giegerich, R., Møller, A.: Analyzing ambiguity of context-free grammars. Sci. Comput. Program. **75**(3), 176–191 (2010)
9. Clark, A., Fox, C., Lappin, S. (eds.): The Handbook of Computational Linguistics and Natural Language Processing. Wiley, Malden (2013)
10. Earley, J.: An efficient context-free parsing algorithm. Commun. ACM **13**(2), 94–102 (1970)
11. Grune, D., Jacobs, C.: Parsing Techniques, a Practical Guide. Monographs in Computer Science, 2nd edn. Springer, New York (2007)
12. Kennedy, K., Ramanathan, J.: A Deterministic Attribute Grammar Evaluator Based on Dynamic Sequencing. ACM Trans. Program. Lang. Syst. **1**(1), 142–160 (1979)
13. Rodriguez-Cerezo, D., Henriques, P.R., Sierra, J.L.: Attribute grammars made easier: EvDebugger a visual debugger for attribute grammars. In: 2014 International Symposium on Computers in Education (SIIE), pp. 23–28. IEEE (2014)
14. Scott, E.: SPPF-style parsing from earley recognisers. Electron. Notes Theor. Comput. Sci. **203**(2), 53–67 (2008)
15. Scott, E., Johnstone, A.: Recognition is not parsing—SPPF-style parsing from cubic recognisers. Sci. Comput. Program. **75**(1), 55–70 (2010)
16. Sierra, J.L., Fernández-Pampillon, A.M., Fernández-Valmayor, A.: An environment for supporting active learning in courses on language processing. ACM SIGCSE Bull. **40**(3), 128–132 (2008)
17. Tomita, M.: Efficient Parsing for Natural Language. Kluwer Academic, Boston (1986)
18. Vasudevan, N., Tratt, L.: Detecting ambiguity in programming language grammars. In: Erwig, M., Paige, R.F., Van Wyk, E. (eds.) SLE 2013. LNCS, vol. 8225, pp. 157–176. Springer, Heidelberg (2013)

Assessing Attribute Grammars' Quality: Metrics and a Tool

João Cruz[1], Pedro Rangel Henriques[1]([✉]), and Daniela da Cruz[2]

[1] Dpt.Informatica/Centro Algoritmi, Universidade do Minho, Braga, Portugal
pedrorangelhenriques@gmail.com
[2] Centro Algoritmi, IPCA, Barcelos, Portugal

Abstract. The definition of metrics and their evaluation process is an activity intrinsic to each engineering branch and it has to do with the need to reason quantitatively about the quality of the developed products. Years ago software engineers working on the field of formal languages and grammars came out with the idea of measuring grammars. However no much progress was done in this trend; there is a clear lack for tools to automatize the computation of some grammar metrics grammars. In this paper we will introduce a tool, GQE, aimed at evaluating a new set of simple metrics for attribute grammars (AG) in order to help on the assessment of AGs quality.

Keywords: Grammar quality · Grammar metrics · Language quality

1 Introduction

Grammar Engineering [1,4,7,8] is a field in software engineering that involves the application of well studied software techniques and methods to grammars, just as they are applied on another software products. Such techniques include version control, static analysis, unit testing, software metrics and evolution, refactoring, among others. Through their implementation, in today's process of developing and maintaining large grammars, better results can be achieved in terms of quality, increasing their efficiency and usability. The objective of this paper is to introduce a software tool, GQE–Grammar Quality Evaluator, that helps on the assessment of grammar quality by performing the automatic evaluation of a large set of metrics. Based on the metrics computed, any Grammar Engineer will easily be able to reason about the quality of its grammar and to improve it. Although there exist some tools similar to the one here described, such as SynQ by Power and Malloy [9] and gMetrics tools [3], the proposed GQE system is different because it is extended to deal with AGs (not only GFGs). Despite using similar procedures, it will produce different results because new metrics will be considered for the assessment (notice that besides the traditional *size* metrics, we contribute with a new set of *style* and *lexicographical* metrics).

In this paper, before introducing the tool, Sect. 4, we discuss our proposal for factors that define the quality of an AG and characteristics that impact on them, Sect. 2, and introduce the metrics that we intended to evaluate aimed at quantifying the quality, Sect. 3.

© Springer International Publishing Switzerland 2015
J.-L. Sierra-Rodríguez et al. (Eds.): SLATE 2015, CCIS 563, pp. 137–144, 2015.
DOI: 10.1007/978-3-319-27653-3_13

2   Grammar Quality

As a grammar is a two fold formalism used to *define (generate) a language*, and *guide the recognition of that language*, Henriques proposed in [5] a set of factors that shall be considered to assess the grammar quality: *(while language generator)* **Usability** of the grammar as a tool for sentences derivation: ease of **understanding**(learning); ease of **derivation**(writing); ease of **maintenance**. *(while program generator)* **Efficiency** of the grammar as a tool for language processors derivation, considering both the **efficient parsing** of the language sentences (obviously the main concern), and the **efficient generation** of the language processors.

Usability in general measures the level of satisfiability of the user following the grammar to use/understand the language. The **understanding** easiness is related to: the identifiers chosen for the non-terminal and terminal symbols and for the attributes; the use of unit productions; the length of the productions right side (RHS); the notation employed to write the grammar rules (pure or extended BNF); the type of recursion used in the derivation rules (right or left, direct or indirect recursion). Concerning the **derivation**, its easiness depends on: the number of non-terminals and keywords; the number of productions; the use of a consistent notation for the productions as well as a regular recursive schema; the use of clear identifiers. Regarding **maintenance**, besides all the factors exposed above, two more elements are important: modularity (a monolithic version is different of one based on the imported components); complexity, as it reflects the way symbols depend on each other.

The **Efficiency in Recognition** is measured in terms of: *Parsing* time; size and complexity of the *Parsing Tables*. The **Efficiency in automatic Generation of the processor** is measured in terms of: the generation time; the size of the intermediate data structures used for storing and transforming the grammar. The **Efficiency** of the generated processor (the parser), or of the generation process, is affected by factors external to the grammar (like the methods, techniques and algorithms used), but it also depends on the size of the grammar and on its writing style.

After proposing the *factors* that determine the quality of a grammar, it is crucial to identify the *grammar characteristics* that have impact on those factors. For the sake of space we just sum up our research listing the grammar characteristics (able to be measured), which we strongly believe that have a directly influence on the grammar quality: the Identifiers of Symbols or Attributes; the number of Symbols or Attributes, of Productions[1] and Unit Productions; the length of the RHS[2]; the Notation and the Recursion schema used to write the Productions; the Attribute Types and simplicity of the Semantic Operators; the number of Semantic Rules (attribute evaluation rules, contextual conditions, and translation rules); the Attributive schema (purely synthesized, or mixed (inherited and synthesized)); the Syntatic/Semantic Complexity (Symbol/Attribute Dependencies); the Modularity.

---

[1] Or Derivation Rules.

[2] Right-Hand Side.

3   Grammar Metrics

Once identified a list of characteristics[3] that should be taken into account to appraise the quality of a grammar, and considering the proposals by other authors cited in the Introduction, we defined a set of parameters (metrics) that can be measured in a objective and systematic way. Below we just list (without rigorous definitions[4]) the metrics so far identified that are evaluated by GQE, separating those that are extracted from the CFG (concerned with the syntax) from those that are related to the semantics, extracted from the AG.

3.1   CFG Metrics

Assuming G is a *well-formed*[5] Context-Free Grammar, and SDG is the respective Symbol Dependency Graph, we define below the metrics to assess the quality of G, dividing them into 3 groups:

– Size Metrics:
  - (SM1) **Grammar size**
  - (SM2) **Grammar syntax complexity**
  - (SM3) **Parser size**
– Style Metrics
  - (FM1) **form of Recursion**
  - (FM2) **type of Recursion**
  - (FM3) **notation**
– Lexicographical Metrics
  - (LM1) **clear identifiers** for terminal and non-terminal symbols
  - (LM2) **clear reserved-words and signs** from the language defined by G
  - (LM3) **flexibility of terminal-classes**
  - (LM4) **comment types**

   Notice that the evaluation of the above described lexicographic metrics (as well as those that will be introduced in the next subsection for attribute identifiers) rely upon the notion of **Identifier Derivation from a Concept name**. We say that an *identifier*[6] *derives from a multi-term concept name* (for instance, the identifier `DestAddrLst` derives from the concept name *Destination Address List*) if, after applying to it the traditional techniques for *identifiers split and expansion* (for complete definition and details on that topic, please see [2]) one can get a correct concept name in the domain of application of the grammar under analysis.

---

[3] That, at the best of our knowledge, is novel.

[4] Please refer to [6] to find details.

[5] For each N exists at least one derivation rule with that symbol on the LHS, and there are not unreachable N.

[6] For a Terminal, a Non-Terminal, an Attribute or an Attributive Operation.

3.2 AG Metrics

Assuming *AG* is a *well-formed* Attribute Grammar and keeping all metrics introduced before for the assessment of the underlying Context-Free Grammar, we present below the 3 groups of metrics to appraise the quality of an AG:

- Size Metrics:
  - (ASM1) **Attribute Grammar size**
  - (ASM2) **Grammar semantic complexity**
- Style Metrics
  - (AFM1) **attributes complexity**
  - (AFM2) **complexity of the attributive operations**
  - (AFM3) **evaluation scheme** for writing CRs
  - (AFM4) **semantic restriction scheme** for writing CCs
  - (AFM5) **translation scheme** for writing TRs
  - (AFM6) **style of the language** to write the attributive operations
  - (AFM7) **language specificity** to write the attributive operations
- Lexicographical Metrics
  - (ALM1) **clear identifiers** for attributes
  - (ALM2) **clear identifiers** for attributive operators.

4   A Tool for Metric Evaluation

GQE - *Grammar Quality Evaluator* is an attribute grammar compiler (processor) written in Java, and generated by AnTLR from the meta-grammar originally designed by Sam Harwell and Terence Parr, and afterwards extended by us with the necessary attributes and semantic rules to perform the computations of the *size, style and lexicographic metrics* we need. So the tool will accept as input CF grammars and Attribute grammars, in the ANTLR version 4.5 format and, as output, will return a list of evaluated grammar metrics. In the future we aim at produce a more elaborated output, providing a grammar quality report, as discussed below.

Concerning metrics evaluation, we can say that: most of the *Size metrics* are evaluated by direct measuring, with the exception of syntax and semantic complexity that are calculated by building both the Local Dependency and the Symbols Dependency Graphs; the computation of *Style metrics* require a higher degree of complexity because it is necessary to work on the internal data structures used to represent the production set in order to detect definition patterns; regarding the *Lexicographic metrics* some external Natural Language Processing tools, such as IdSplitter and WordNet (among others) were used to analyze the clarity of identifiers.

Some of the metrics (concerned with the grammar style and language style and specificity) can, at a first insight, appear to be useless for grammars written in AnTLR format, but in the future, when GQE is adapted to accept other kind of grammar formats, such metrics will give useful information regarding the grammar quality.

4.1   GQE Results

At this point we are already able to show some results of our evaluation system. It was tested with small, medium and large size grammars (for example, the C Language Context-Free Grammar that has 239 productions) and had a good performance, producing the expected output metrics. In this paper, first, we will present the output evaluated by GQE for a small CFG designed to specify the Lisp language (Fig. 1), and next the same grammar but improved with Attributes, so we will cover all the produced results while explaining some of the metrics implementation and why some metrics are impossible to evaluate in the AnTLR grammar format. Other details on GQE, input and output formats, results, etc., can be seen at http://www.di.uminho.pt/~gepl/GQE.

```
grammar Lisp;

lisp : sExp ;
sExp : NUM
     | WRD
     | '(' sExpList ')';
sExpList : sExp sExpList
         |
         ;
```

Size Metrics	
#T	4
#N	3
#P	6
#UP	0
#R	2
§RHS	1,3
§RHS-Max	3
§Alt	2,0
§Alt-Max	3
#Mod	0
FanIn	2,7
FanOut	1,1
§RD	7
§TabLL	15
§AD-LR	10
§TabsLR	50;30

Fig. 1. Context-Free Grammar of the Lisp language, written in AnTLR format.

Fig. 2. Size Metrics outputted by GQE for the Lisp grammar of Fig. 1

All the size metrics shown in (Fig. 2) were computed by GQE, even the most complex such as FanIn/FanOut(Grammar Syntax Complexity) and those regarding the Parser Size. The same can be said about all the style metrics presented in (Fig. 3) which involve complex pattern matching algorithms. Concerning lexicographic metrics, they were also evaluated automatically with the help of the IdSplitter, Identifier Splitter Expander tool that is able to identify clear identifiers and comment types. The other two metrics: clear reserved-words and signs and terminal-classes flexibility can not be evaluated automatically for obvious reasons (Fig. 4).

Style Metrics	
Recursion Form	FormMixedRec
Recursion Type	Right Recursion
Notation	Pure-BNF

Fig. 3. Style Metrics results from GQE for the input lisp grammar of Fig. 1

Identifier	Splitter/Expansion	Lexicographic Metrics	
lisp	lisp	Clear Identifiers	5 / 5
sExpList	symbolic / expression / list	Clear Keywords/ Signs	?
sExp	symbolic / expression	Flexibility Terminal-Classes	?
NUM	number		
WRD	word	Comment Type	0

Fig. 4. Lexicographic Metrics results from GQE for the input lisp grammar of Fig. 1

Size Metrics	
#A	17
#AI	8
#AS	9
#CR	23
#CC	*
#CR	*
FanIn	0,81
FanOut	1,14

Style Metrics	
#Att. Complexity	5 / 17
#Att. Op. Complexity	?
Evaluation Scheme	?
Semantic Restriction Scheme	*
Translation Scheme	*
Language Style	Imperative
Language Specificity	Standard OO

Fig. 5. Size Metrics outputted by GQE for the input Lisp AG.

Fig. 6. Style Metrics outputted by GQE for the input Lisp AG.

Now, to show the tool results for attribute grammars, we used the grammar of Fig. 1, and add some attributes and some semantic information, the source file of that grammar can be seen at the link refereed before.

To be possible to produce those results for AGs, the proposed metrics implementation complexity is bigger. Size metrics are shown in (Fig. 5), it is important to say that for AGs written in AnTLR format it is impossible to evaluate the number of Transition Rules and Context-Conditions, because they are not specified explicitly, in Anltr.

The same argument works for the metrics Semantic Restriction Scheme and Translation Scheme that belong to the set of style metrics in Fig. 6; for the Evaluation Scheme and Attributive Operation Complexity metrics, they are being implemented and need some more work and test.

Lexicographic Metrics	
Clear Attributes Identifiers	17 / 17
Clear Attributive operators Identifiers	?

Fig. 7. Lexicographic Metrics outputted by GQE for the Lisp grammar.

Finally, for the lexicographic metrics (Fig. 7), once more the Splitter Expander helps to identify clear attribute identifiers; attributive operator identifier metrics require some more human opinion, our idea is that the grammar engineer works together with GQE.

5 Conclusion

We have introduced GQE, a tool to support the user in assessing Attribute Grammars. The motivation to start this research project was given in the Introduction; it mainly arose from the need to assess the quality of software products and specifications using quantitative measurements. GQE computes a set of fine grain metrics, built upon the traditional metrics for CFGs and extended with new metrics that consider the writing style (syntactic recursive schemas, attribute evaluation schemas, notation, etc.) and also the understandability inherent to the identifers. The tool (developed in Java with the help of AnTLR) reads any AnTLR Grammar (this is, any AG written in the AnTLR metalanguage) and outputs the value for each one of the metrics under consideration. The user (for sure a Grammar Engineer) will analyze the values provided and will be able to come up with an assessment. Easily he will be able to transform his original grammar and submit the new one for re-evaluation to understand the eventual improvements. We strongly believe that this approach will help the user in designing equivalent grammar versions (i.e., different grammars that generate precisely the same language) with different quality levels according to different perspectives (enhancing the grammar's usability or its efficiency). Instead of producing a final figure quantifying (or grading) the grammar's quality, we intend to create a database (a kind of CBR system) to collect grammars, metric values computed, and the expert's grading, in order to use machine learning algorithms to infer the grading in future cases. This will enable us to provide a more elaborated report on the grammar quality and suggest a set of possible transformations to improve the grammar (according to different perspectives).

As future work we also plan to related the grammar with the generated language in order to understand until which point is it possible to discuss also the language quality.

Acknowledgment. This work is co-funded by the North Portugal Regional Operational Programme, under the National Strategic Reference Framework (NSFR), through the European Regional Development Fund (ERDF), within project GreenSSCM - NORTE-07-02-FEDER-038973.

References

1. Alves, T.L., Visser, J.: A case study in grammar engineering. In: Gašević, D., Lämmel, R., Van Wyk, E. (eds.) SLE 2008. LNCS, vol. 5452, pp. 285–304. Springer, Heidelberg (2009)
2. Carvalho, N.R., Almeida, J.J., Henriques, P.R., Pereira, M.J.V.: From source code identifiers to natural language terms. J. Syst. Softw. **100**, 117–128 (2015). http://dx.doi.org/10.1016/j.jss.2014.10.013
3. Crepinsek, M., Kosar, T., Mernik, M., Cervelle, J., Forax, R., Roussel, G.: On automata and language based grammar metrics. Comput. Sci. Inf. Syst. **7**(2), 309–329 (2010). http://dx.doi.org/10.2298/CSIS1002309C
4. Erbach, G.: Tools for grammar engineering. In: Proceedings of the Third Conference on Applied Natural Language Processing, ANLC 1992, pp. 243–244. Association for Computational Linguistics, Stroudsburg (1992)
5. Henriques, P.R.: Brincando às Linguagens com Rigor: Engenharia Gramatical. Technical report, Dep. de Informática, E.Engenharia da Universidade do Minho, October 2011. habilitation monography presented and discussed in a public session held in April 2012
6. Cruz, J.: An Attribute Grammar based System to assess Grammars Quality (PreThesis) (2015)
7. Klint, P., Lämmel, R., Verhoef, C.: Toward an engineering discipline for grammarware. ACM Trans. Softw. Eng. Methodol. **14**(3), 331–380 (2005). http://doi.acm.org/10.1145/1072997.1073000
8. Lämmel, R.: Grammar testing. In: Hussmann, H. (ed.) FASE 2001. LNCS, vol. 2029, pp. 201–216. Springer, Heidelberg (2001). http://dx.doi.org/10.1007/3-540-45314-815
9. Power, J.F., Malloy, B.A.: A metrics suite for grammar-based software. J. Softw. Maintenance **16**(6), 405–426 (2004). http://dx.doi.org/10.1002/smr.293

A Syntax-Directed Model Transformation Framework Based on Attribute Grammars

Antonio Sarasa-Cabezuelo and José-Luis Sierra[✉]

Facultad de Informática, Universidad Complutense de Madrid, Madrid, Spain
{asarasa, jlsierra}@fdi.ucm.es

Abstract. Model transformation is a key aspect of model-driven software development because it enables the automatic derivation of different interpretations of a system model. In many scenarios (e.g., design of domain-specific languages), models usually have implicit identifiable primary tree-like syntactic structures, on which additional secondary relationships are imposed to yield the final model graphs. Therefore, in these scenarios it seems natural to address the processing of these models on the basis of their underlying syntactic structure. For this purpose, we have developed AGT, an experimental transformation framework based on attribute grammars, which takes full advantage of the underlying syntactic structure of source models. For models in which this structure is clearly identifiable, the approach could result more natural and easier to use and maintain than other more conventional model transformation approaches (e.g., those based on more standard model transformation languages).

Keywords: Attribute grammar · Model-driven development · Model transformation

1 Introduction

A key aspect of model-driven software development is model transformation, i.e. the translation of models conforming a particular meta-model (source meta-model) into models conforming another meta-model (target meta-model).

This paper focuses on formal grammars for model transformation, specifically *attribute grammars* [3]. For this purpose it describes a Java framework for model transformations called AGT (*Attribute Grammar Transformer*), which enables specifications based on attribute grammars to describe model to model transformations in a declarative style. Contrarily to the main trend in the grammar-based approach, the use of *graph grammars* [2], attribute grammars enable syntax-directed transformation processes organized around the tree-like primary structure of the model, while graph grammars adopt a template-based approach, in which subgraphs are matched and transformations applied on the matched subgraphs. Therefore, attribute grammars can be more appropriate for models with a well-distinguished hierarchical structure (e.g., models arising in the modelling of domain-specific languages). In addition, contrarily to other works of using attribute grammars for model transformation, like [1], the proposal described in this paper is not based on canonical textual encodings of the models, but it adapts the attribute grammar formalism to work directly on object

© Springer International Publishing Switzerland 2015
J.-L. Sierra-Rodríguez et al. (Eds.): SLATE 2015, CCIS 563, pp. 145–152, 2015.
DOI: 10.1007/978-3-319-27653-3_14

networks (and, therefore, on the in-memory representation of models). By doing it so, the hierarchical primary structure of models is fully exploited to guiding the transformation process in a syntax-directed fashion.

The rest of the paper is structured as follows. Section 2 briefly introduces attribute grammars. Section 3 gives an overview of AGT. Section 4 describes AGTL, the transformation specification language included in AGT. Finally, Sect. 5 outlines some conclusions and lines of future work.

2 Attribute Grammars

Attribute grammars is a formalism that allows the description of the syntax and the translational semantics of programming languages [3]. The syntax is described using context-free grammars and the semantics are described using semantic attributes associated with the grammar symbols and semantic equations associated to each grammar rule.

Semantic attributes can be synthesized or inherited, and they take values in the nodes of the parse trees associated to the sentences of the language. Each equation describes relations between attributes that indicate how the values of the synthesized attributes in the rule LHS and the inherited attributes in the rule RHS are computed by applying semantic functions on the attributes used in this calculation.

A key property of the attribute grammars concerns the order of application of the semantic equations. In this sense, it is not necessary to explicitly specify such an order i.e., the *evaluation order* of the attributes, because it is derivable from the dependencies among attributes imposed by the equations. It makes of attribute grammars a more declarative and high-level formalism than other approaches to syntax-directed translation specification (e.g., *translation schemata*).

3 The AGT Framework

AGT (*Attribute Grammar Transformer*) is a model transformation framework based on attribute grammars implemented and tightly integrated with the Java language. Indeed, the framework is oriented to transform (or, more generally, to process) models encoded as Java object networks (in this way, meta-models are mirrored on sets of Java classes). This framework specializes Java by providing the following main components: (i) AGLT (*Attribute Grammar Transformation Language*), a declarative specification language for attribute grammar-based model transformations, (ii) AGT transformer, a transformation engine that implements the operational semantics of AGTL, and (iii) AGLT runtime, a set of Java utility classes that can be used during transformation.

Transformations according to AGT actually integrate two well-differentiated parts: (i) on one hand, a syntax-directed specification provided as an AGTL attribute grammar, and (ii) on another hand, additional Java code implementing the specific machinery required to carry out the implementation.

The connection between the two parts is given by a *semantic class*, which implements the semantic functions used in the AGLT specification in terms of the additional Java machinery provided. This organization is analogous to that followed in other syntax-directed approaches to information processing developed in our group (e.g., XLOP for XML processing [5] or JLOP [4] for JSON processing).

4 AGTL: The AGT Specification Language

The core component of AGT is AGTL (*Attribute Grammar Transformation Language*), a specification language for attribute grammar-based model transformations. This section summarize the AGTL concrete syntax (Subsect. 4.1), the AGTL abstract syntax (Subsect. 4.2) and the AGTL operational semantics (Subsect. 4.3), and it gives a small example of using AGTL (Subsect. 4.4).

4.1 AGLT Concrete Syntax

A specification in AGTL is composed by (see Fig. 1): (i) a declaration of non-terminal symbols and their associated attributes, and (ii) a specification of the attribute grammar rules. Each rule includes a context-free part connecting non-terminals with classes of the source model, and semantic equations specifying how to compute the required attributes in the rule context.

```
<Grammar> ::=  <NTDec>+ <Rule>+
<NTDec>   ::= nt(nonTerminalName,[<AttList>],[<AttList>]).
<AttList> ::= (attName (, attName)*)?
<Rule> ::= nonTerminalName '::='
              (className {<FieldSpecList>}
                     {<EquationList>}
                     ({<EquationList>})? |
                 null {<EquationList>}          |
                className: nonTerminalName {<EquationList>}) .
<FieldSpecList> ::= (<FieldSpec> (, <FieldSpec>)*)?
<FieldSpec> ::= fieldName : nonTerminalName
<EquationList> ::= (<Equation> (, <Equation>)*)?
<Equation> ::= <AttrRef> = <SemExp>
<AttrRef> ::= nonTerminalName( '(' num ')' )?.attname
<SemExp> ::= <AttrRef>|funName '(' <ArgList> ')'|@fieldName
<ArgList> ::= (<SemExp> (,<SempExp>)*)?
```

Fig. 1. Textual syntax of the AGTL language

Each non-terminal has associated rules in the AGTL grammar. AGTL distinguishes between three different types of rules:

- *Class* rules. Rules applicable to objects of a given class. A class rule specifies such a class of objects to which the rule can be applied, a *field specification* (see below), and, in addition to the set of *first visit* equations, a set of *cut* equations.
- *Null* rules. Rules applicable to *null* values.

- *Bridge* rules. Rules used to forward the processing of an object to a more specific rule depending on its class. Thus, these rules are a simple but yet effective way of dealing with inheritance in the source model. The forwarding is carried out by providing a non-terminal name.

Concerning field specifications in *class* rules, each field specification makes it possible to select the rule to be applied in order to analyze a field of the matched object. The rule to apply is selected by providing a non-terminal name.

Concerning semantic equations, more representative AGTL rules include two different types of equations:

- *First-visit* equations. These equations are used the first time the rule is applied to an object.
- *Cut* equations. These equations are used when the rule is applied to an already-analyzed object.

This organization of semantic equations in *first-visit* and *cut* packages is necessary to enable the application of AGLT to arbitrary object graphs.

Concerning semantic expressions used in semantic equations, AGTL allows two types of basic expressions: (i) references to attributes in the rule, and (ii) field value extraction expression, which make it possible to query the values of fields in objects to which the rules are applied (the operator @ is used on field names). In this way, it makes apparent how semantic attributes in AGLT do not correspond to class fields, but they represent placeholders in which to store transformation results, as it is usual with attribute grammars.

Finally, compound expressions are formed by applying semantic functions to simpler expressions.

4.2   AGTL Abstract Syntax

The AGTL abstract syntax is given by a set of classes that represent the different concepts of the language. Figure 2 summarizes this abstract syntax.

The `Grammar` class represents whole AGLT specifications. The `NonTerminalDescription` class represents descriptions of non-terminals. `synAttributes` and `inhAttributes` lists in `NonTerminalDescription` allows for the description of synthetized and inherited attributes.

The `Rule` abstract class represents AGLT rules. The non-terminal to which the rule is associated is given by the `lhs` field. In addition, the `firstVisitEquations` list represents the list of conventional first-visit semantic equations associated to the rule. This class is, in its turn, specialized by one concrete class for each type of rule envisioned by AGLT:

- *Class* rules are modeled by the `ClassRule` class. The class of objects to which the rule can be applied is given by `theClass` field. The field specification is modelled by the `FieldSpecification` class (`field` gives the name of the field and `nonTerminalName` the name of the non-terminal used to index the rule to apply). Finally, the `cutEquations` list represents the list of cut equations associated to the class rule.

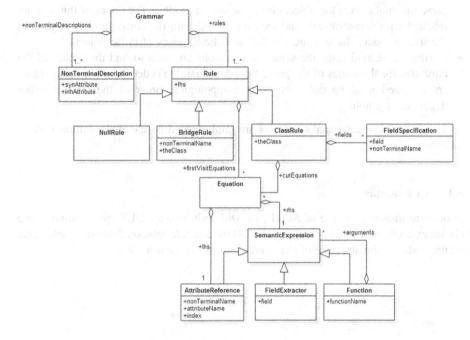

Fig. 2. AGTL abstract syntax

- *Null* rules are modeled by the `NullRule` class.
- *Bridge* rules are modelled by the `BridgeRule` class. Here `theclass` identifies the class of objects to which the rule is applied and the `nonTerminalName` binds to the non-terminal used to forward the processing.

Semantic equations are modelled by the `Equation` class. In this class `lhs` represents the equation's left-hand side, and `rhs` the equation's right-hand side.

Attribute references are modelled by the `AttributeReference` class (the name of the attribute is given by `attributeName`, the name of the non-terminal by `nonTerminalName`, and the occurrence number of the non-terminal in the rule by `index`).

Finally, semantic expressions are modelled by the `SemanticExpression` class. The `FieldExtraction` class models field value extraction expressions, and the `Function` class models compound expressions (the semantic function applied is represented by the `functionName` field).

4.3 AGTL Operational Semantics

With respect to operational semantics, AGTL follows a syntax-directed translation model. More particularly, transformation is organized in two different stages:

- During the first stage, the *analysis stage*, syntax rules are applied in order to make the syntactic structure of the source model explicit. This structure is represented as a parse tree that covers the objects from the source models. The nodes of this tree are labelled with non-terminals and they are linked with the corresponding objects in the source model. Its arcs are labelled with field names of those objects.

- During the second stage the semantic equations are used to find the values of the attributes for the nodes of the parse tree. As usual, the order in which these values are obtained must be consistent with a topological order of the corresponding dependency graph.

In AGT these operational semantics are implemented by the aforementioned AGT transformer.

4.4   An Example

In order to illustrate the use of AGTL, we will address the AGLT specification of the classic example of a simple mapping of class models into relational database schemata. Figure 3 shows the structure of the source and target meta-models.

(a)   (b)

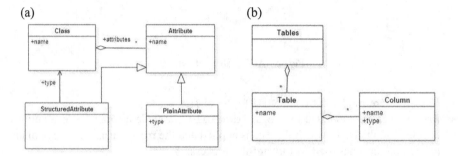

Fig. 3. (a) Source metamodel; (b) Target metamodel

The transformation will map each class in the class model into a table in the relational schema. The table name will match the class name. Columns names will match attribute names in the classes. The types of the columns will depend on the type of attributes, so the type of the column will be the type of the attribute if it is a primitive type (`boolean`, `integer`, `char`) or "fk "(foreign key) followed by the class name if the type of the attribute is another class.

Figure 4 shows the transformation specified in AGTL. In the specification, the inherited attribute `itables` is intended to contain the set of tables already constructed, and the synthetized attribute `tables` the overall set of tables. The class `com.agt.core.Array` is a utility class of the AGT runtime that is useful in order to deal with collections (the AGT transformer automatically wraps collections using this utility class). Objects of this class exhibits two fields (`first`: the first element of the collection, `butfirst`: a collection made of the rest of fields). The implementation of the semantic functions used must be provided to the AGT transformer as a plain Java class (the semantic class of this transformation) In addition, it is also needed to indicate

```
nt(<Root>,[],[tables]).
nt(<Class>,[itables],[tables,name]).
nt(<Attributes>,[itables],[tables,cols]).
nt(<Attribute>,[itables],[tables,col]).

<Root> ::= Class: <Class> {
                <Class>.itables =  mkEmptySetOfTables(),
                <Root>.tables = mkTables(<Class>.tables)}.

<Class> ::= Class { attributes: <Attributes>   }
    {
        <Attributes>.itables =
           addTable(<Class>.itables,
                    mkTable(@name,<Attributes>.cols)),
        <Class>.tables = <Attributes>.tables,
        <Class>.name = @name}
    {
        <Class>.tables = <Class>.itables,
        <Class>.name = @name}.

<Attributes> ::= com.agt.core.Array {first: <Attribute>,
                                     butfirst: <Attributes>} {
        <Attribute>.itables = <Attributes>(0).itables,
        <Attributes>(1).itables = <Attribute>.tables,
        <Attributes>(0).tables = <Attributes>(1).tables,
        <Attributes>(0).cols = addCol(<Attributes>(1).cols,
                                      <Attribute>.col)}.

<Attributes> ::= null {
        <Attributes>.tables = <Attributes>.itables,
        <Attributes>.cols = mkEmptyListOfCols()}.

<Attribute> ::= PlainAttribute {}{
        <Attribute>.tables = <Attribute>.itables,
        <Attribute>.col = mkCol(@name,@type)}.

<Attribute> ::= StructuredAttribute {type: <Class>}{
        <Class>.itables = <Attribute>.itables,
        <Attribute>.tables = <Class>.tables,
        <Attribute>.col = mkCol(@name, mkForeignKey(<Class>.name))}.
```

Fig. 4. Example of transformation in AGTL

to this transformer: (i) a root object in the source model, and (ii) a root non-terminal in the grammar. Then, the transformation will proceed according the AGTL operational semantics outlined in the previous subsection.

5   Conclusions and Future Work

This paper has presented a model transformation framework based on attribute grammars, which is highly integrated in Java. The aim of the framework is similar to the presented in [1], but differs in that attribute grammars operate on object networks instead of on textual encodings of the source models. It makes it possible to exploit the primary hierarchical structure of models in a more natural way. Indeed, the approach can be particularly well-suited for source models with well-distinguished primary tree-like structures (like those arising, for instance, when modeling domain-specific languages).

We are currently working on the static typing of AGLT specifications. As future lines of work we envision the integration of AGT with standard metamodeling proposals (e.g., MOF or Ecore), the creation of an IDE for AGT, the efficiency analysis of the AGT transformer, and an in-depth empirical evaluation of our proposal. This evaluation will include a comparison with other proposals on a representative workbench of examples, as well as an empirical study concerning usability and efficacy with developers.

Acknowledgements. We would like to thank Juan-Pablo Gracia-Benitez by contributing to a preliminary implementation of the framework. This work has been partially supported by the BBVA Foundation (research grant HUM14_251), by the Spanish R&D&I Plan (research grant TIN2014-52010-R), and by Santander-UCM GR3/14 (group number 962022).

References

1. Dehayni, M., Féraud, L.: An approach of model transformation based on attribute grammars. In: Masood, A., Léonard, M., Pigneur, Y., Patel, S. (eds.) OOIS 2003. LNCS, vol. 2817, pp. 412–423. Springer, Heidelberg (2003)
2. Rozenberg, G. (ed.): Handbook of graph grammars and computing by graph transformation: volume I. foundations. World Scientific Publishing Co., River Edge (1997)
3. Paakki, J.: Attribute grammar paradigms - a high-level methodology in language implementation. ACM Comput. Surv. **27**(2), 196–255 (1995)
4. Sarasa-Cabezuelo, A., Sierra, J.L.: Grammar-driven development of JSON processing applications. In: FedCSIS 2013, pp. 1545–1552 (2013)
5. Sarasa-Cabezuelo, A., Sierra, J.L.: The grammatical approach: a syntax-directed declarative specification method for XML processing tasks. Comput. Stand. Interfaces **35**(1), 114–131 (2013)

An AST-based Tool, Spector, for Plagiarism Detection: The Approach, Functionality, and Implementation

Vítor T. Martins, Pedro Rangel Henriques$^{(\boxtimes)}$, and Daniela da Cruz

Departamento de Informática/Centro Algoritmi,
Universidade do Minho, 4710-057 Braga, Portugal
{vtiagovm,pedrorangelhenriques,danieladacruz}@gmail.com
http://www.di.uminho.pt/eng/

Abstract. We propose a methodology using abstract syntax trees for the detection of plagiarism in source code, within an academic environment.

We show the architecture and decisions that came before we produce our own solution (Spector), after conducting a study of the methods and tools in existence. An example is then shown, which goes through and explains each of the algorithms steps.

Finally, conclusions are drawn noting that such a system, while not the most efficient, produces accurate results.

Keywords: Software · Plagiarism · Detection · Comparison · Test

1   Introduction

In our previous work [3], we discuss the results of our search for source code plagiarism detection methodologies and tools. We found several candidates but, after testing their accuracy by using files modified to hide plagiarism, we saw that most solutions had trouble with some cases.

Given the desire to make detections upon programs and our experience in language engineering, we were motivated to use a compilation aproach. After some research and discussion, we chose to use Abstract Syntax Trees (ASTs) since, by abstracting source code, we can analyze its functionality. Such tools exist and were said to be accurate [1,2] but were not available for download or use. To counteract that lack, our contribution will be the development of one such tool and make it openly available.

2   Approach

Figure 1 shows the representation of an AST generated from a source code. We can see that the original source code defines a function (sum) that adds two integers and uses it to print an equation that adds 4 to 7.

© Springer International Publishing Switzerland 2015
J.-L. Sierra-Rodríguez et al. (Eds.): SLATE 2015, CCIS 563, pp. 153–159, 2015.
DOI: 10.1007/978-3-319-27653-3_15

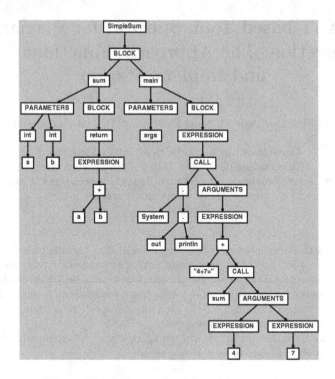

Fig. 1. An *AST* generated from a source code

Knowing that this representation gives us an ordered network of nodes, while retaining the information about their type and contents, it is easy to see that it provides enough information for source code comparison.

In order to facilitate the generation of *ASTs*, we have chosen to use *ANTLR*[1] [4] which allows us to generate parsers from grammar files, which can translate source code into an *AST* automatically.

2.1 Target Characteristics

As previously said, by using an intermediate abstraction, we can easily ignore specific characteristics. So we will focus on those that can be modified without altering the source code functionality. Namely Identifiers (the names of classes, variables, etc.), Expression elements (for ex.: $x, >, 1$), Conditionals (like if and while) and Blocks (a group of statements enclosed between {}).

This means we will ignore other characteristics like comments, as we are focusing on the source codes functionality. For example, if we wanted to compare Identifiers we do not need to check their scope or type, we can simply see if they are used in the same places and have similar behaviors.

---

[1] ANother Tool for Language Recognition.

2.2   Architecture

After making our decisions on how our system would work, we gave it the name
of *Spector* (Source insPECTOR). We also produced a diagram (in Fig. 2) that
shows the various parts and how they relate.

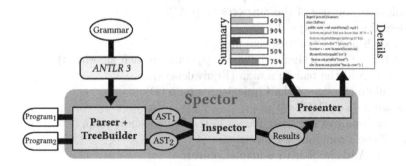

Fig. 2. A diagram of the interaction between the systems parts

In Fig. 2 we can observe that given a Grammar G, we can produce a
Parser+TreeBuilder TB, using *ANTLR*. This TB can then be used to translate a Program P into an AST. Given a pair of Programs (let us say P_1 and
P_2), a TB will be used to produce AST_1 and AST_2. These $ASTs$ will then
be delivered to an Inspector which will produce measures and send them to a
Presenter which will output the results. To support a new language, we would
simply replace G and generate a new TB.

3   Functionality

The functionality is split into 5 algorithms that each produce a measure, along
with a final one that calculates the global measure.

Given source codes S_1 and S_2, we first check if they are not exact copies by
comparing: the number of nodes, the number of nodes by category[2] and finally,
the contents of every node. When all those comparisons match, we return a
measure of 100 %, which avoids using the other algorithms.

For other cases, we have algorithms for each target (listed in Sect. 2.1). While
these algorithms work in different ways, they follow a base functionality:

1. Build a map per source, associating elements to their occurrences (M_1, M_2),
2. Calculate the highest number of items between both Ms (A),
3. For each pair of items between M_1 and M_2,
   (a) If their number of occurrences is similar,

---

[2] In grammatical terms, these categories are the terminals that were assigned to integers by ANTLR (in a tokens file).

      i. Add the pair to a map (Candidates),
4. Calculate the size of the Candidates map (B),
5. Measure $+= (B/A) * W_1$,
6. For each pair in the Candidates map,
    (a) If their number of occurrences by category is similar,
      i. Add the pair to a map (Suspects),
7. Calculate the size of the Suspects map (C),
8. Measure $+= (C/B) * W_2$,
9. For each pair in the Suspects map,
    (a) If they have a similar behavior (specific to each algorithm),
      i. Add the pair to a map (Equivalences),
10. Calculate the size of the Equivalences map (D),
11. Measure $+= (D/C) * W_3$,
12. For each pair of identifier names in the Equivalences map,
    (a) If their contents are the same,
      i. Add the pair to a map (Copies),
13. Calculate the size of the Copies map (E),
14. Measure $+= (E/D) * W_4$,
15. Return Measure.

The W_i variables indicate weight constants and were given a value of 0.18, 0.42, 0.38 and 0.02, respectively. We chose these values based on a few tests. However, they must be adjusted through the use of further tests.

Of course, each algorithm is targeting something different, so they have the following differences:

Algorithm that detects Identifiers: The map (M) associates an Identifier name to its Occurrence nodes (IOM), which is similar if their parent nodes have the same category and the (non-identifier) neighbors[3] have the same contents.

Algorithm that detects Expression elements: The map (M) associates an Expression element to its Occurrence nodes (EOM), which is similar to those whose (non-identifier) neighbors contents are equal.

Algorithm that detects Conditionals: The map (M) associates a Conditional to its Condition node (CCM), which is similar to those whose condition has (non-identifier) nodes with the same contents.

Algorithm that detects Blocks: In this case, two maps are created: one associates a Block node to a Name[4] (BNM) and the other associates each Block node to all of its Children (BCM). The children are: every node inside the block along with the nodes from called blocks. This is done by checking if nodes are calls to internal methods, in which case the contents of the called block are added.

---

[3] The other children of this nodes parent.
[4] This name is the identifier of the parent block, in other words, "$main_{block}$" has a block with "main" as its name.

Main Algorithm: This algorithm calculates a final measure from those produced by the previous algorithms. That final measure is a similarity measure of a pair of Suspects.

Let us consider that each algorithm was implemented in a method $Method_i$, where i is a number from 1 to 5. With $Method_1$ being the one which determines if two source codes are exact copies. The algorithm works as follows:

1. X = Array with 5 elements,
2. for each method i,
   (a) $X[i] = Method_i(S_A, S_B)$,
3. If X[1] is different from 0,
   (a) Return X[1].
4. Otherwise
   (a) Calculate the number of Xs from 2 to 5 that are not 0 (A),
   (b) Measure $= \left(\frac{X[2]+X[3]+X[4]+X[5]}{A} \right) * 100$,
   (c) Return Measure.

As we can see, the algorithm either returns the X[1] measure (which is either 0 % or 100 %) or the average computation done using the other results (X[2] to X[5]) that were not 0 %.

Threshold. Since the algorithms match the number of occurrences when checking if the elements should be added to a Candidates map, the comparisons will be limited to cases with an equal number of occurrences. Which led us to the addition of a similarity threshold, which specifies the strictness of the comparisons. As an example: If we were comparing the number of nodes within two blocks and the first had 10 nodes, a threshold of 20 % means that the second *AST* must have between 8 and 12 nodes to be considered similar.

4   Implementation

To keep *Spector* modular, we split its functionality into two packages. A **lang** package which contains a Suspect class that will have the input generated from a source code and, for each language: The Parser+TreeBuilder classes and a Nexus class which interfaces with them. Along with a **spector** package that contains the main classes (Spector, Inspector and Presenter) along with their auxiliary classes (FileHandler and Comparison).

4.1   Features

We list below the main features provided by our tool:

1. Can output summary and/or detailed results,
2. Accepts submissions as groups of files,
3. Works offline,
4. Available as Open Source.

It is important to note that, the first two are relevant for any plagiarism detector and the other two are crucial since we want the tool to be available for integration into other systems.

5   Example

A complete example comparing 2 similar files (AllIn1 and AllIn2), from a semantic point of view, can be found with a detailed description at the following website: http://www.di.uminho.pt/~gepl/Spector/paper/slate15/examples/algorithms.pdf

Notice that the example finishes with the following computation:

$$\left(\frac{0.941 + 0.941 + 0.98 + 0}{3}\right) * 100 = 95.4\,\%$$

This gives us a similarity measure of 95.4 %, which indicates that the source codes are very similar. If we use the tool, produced following our methodology, to compare this test, we would read a different result in the HTML summary outputed and reproduced in Fig. 3.

Spector

First File	Second File	Similarity
AllIn1.java	AllIn2.java	72.242%

Fig. 3. The HTML produced, showing the similarity measure.

As we can see, instead of being close to the expected 95.4 %, the result is instead 72.242 %. Detailed results show that the calculation was instead:

$$\left(\frac{92.45 + 94.017 + 98 + 4.5}{4}\right) = 72.242\,\%$$

This was due to the last algorithm returning a 4.5 as its similarity measure. The algorithms details show that the tool mistakingly associated the **AllIn1** to the **main** block in the Candidates map, due to them having the same number of child nodes. This shows us the importance of the weights in the algorithms since we want to avoid such false negatives as well as false positives.

6   Conclusion

In this paper, we have discussed our approach on building a tool that will detect plagiarism in source code named Spector. We have defined its structure and the decisions that were behind it. We have also seen the algorithms that will drive the comparisons and how they are used together to produce similarity measures. Seeing as we focused on detecting similarity between source code structures, the resulting tool is a *source code similarity detector* and is likely to report false-positives when faced with smaller code.

As future work we have the improvement of the results in terms of information about the associations established. Along with the upgrade of the Java grammar, to support the latest version of the Java language and the extension with grammars to cope with other languages.

The next step will be to perfect the tools implementation and test it against bigger test cases so that an optimal threshold and weights may be determined.

Acknowledgments. This work is co-funded by the North Portugal Regional Operational Programme, under the National Strategic Reference Framework (NSFR), through the European Regional Development Fund (ERDF), within project GreenSSCM - NORTE-07-02-FEDER-038973.

References

1. Bahtiyar, M.Y.: JClone: syntax tree based clone detection for Java. Master's thesis, Linnæus University (2010)
2. Cui, B., Li, J., Guo, T., Wang, J., Ma, D.: Code comparison system based on abstract syntax tree. In: 2010 3rd IEEE International Conference on Broadband Network and Multimedia Technology (IC-BNMT), pp. 668–673 (2010)
3. Martins, V.T., Fonte, D., Henriques, P.R., da Cruz, D.: Plagiarism detection: a tool survey and comparison. In: Pereira, M.J.V., Leal, J.P., Simões, A. (eds.) 3rd SLATE. OASIcs, vol. 38, pp. 143–158. Schloss Dagstuhl-Leibniz-Zentrum fuer Informatik, Dagstuhl, Germany (2014). http://drops.dagstuhl.de/opus/volltexte/2014/4566
4. Parr, T.J., Quong, R.W.: ANTLR: a predicated-LL(k) parser generator. Softw.-Pract. Experience **25**(7), 789–810 (1995). http://citeseerx.ist.psu.edu/viewdoc/summary?doi=10.1.1.15.70

Towards the Generation of Graphical Modelling Environments Aided by Patterns

Antonio Garmendia$^{(\boxtimes)}$, Ana Pescador, Esther Guerra, and Juan de Lara

Modelling and Software Engineering Research Group,
Computer Science Department, Universidad Autónoma de Madrid, Madrid, Spain
antonio.garmendia@uam.es
http://miso.es

Abstract. Model-Driven Engineering (MDE) promotes the use of models to conduct all phases of software development in an automated way. Such models are described using Domain Specific Modelling Languages (DSMLs). While the definition of DSMLs and their supporting environments are recurring activities in MDE, they are mostly developed ad-hoc from scratch. This paper proposes the use of patterns to describe the abstract and concrete graphical syntax of DSMLs, and to automate the generation of a graphical modelling environment for them.

Keywords: Model-Driven Engineering · Domain specific languages · Patterns · Graphical modelling environments

1 Introduction

Model-Driven Engineering (MDE) promotes a model-centric approach for software development, where models are used to specify, design, test, simulate and generate code for applications. While models can be described using general-purpose modelling languages, like UML, it is frequent the use of Domain Specific Modelling Languages (DSMLs) focussed on the particularities of a domain [8].

Hence, the creation of DSMLs is recurrent in MDE, for which one needs to describe their abstract and concrete syntax, their semantics, and developing a suitable modelling environment for them. Although there are software frameworks to ease the development of textual and graphical environments [8,10,12], the creation of DSMLs is mostly an ad-hoc process lacking the ability to build on existing knowledge coming from the construction of similar DSMLs.

To simplify the creation of DSMLs, we propose their assisted construction by means of patterns. In particular, *domain patterns* describe recurring concepts common to a domain, and *concrete syntax patterns* gather standard representation options for DSMLs and enable the synthesis of modelling environments. As a proof of concept, we show a prototype implementation for Eclipse.

The remaining of the paper is organized as follows. First, Sect. 2 introduces our approach. Then, Sect. 3 explains how to build graphical DSMLs with patterns. Next, Sect. 4 reviews related works, and finally, Sect. 5 concludes.

J.-L. Sierra-Rodríguez et al. (Eds.): SLATE 2015, CCIS 563, pp. 160–168, 2015.
DOI: 10.1007/978-3-319-27653-3_16

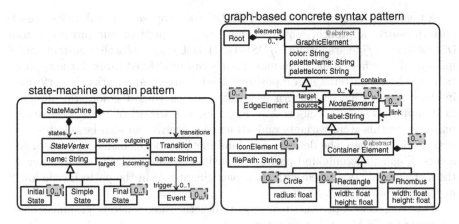

Fig. 1. Domain pattern (left). Graph-based concrete syntax pattern (right).

2   Overview

The design of a DSML encompasses several aspects, including abstract syntax, concrete syntax, and semantics. In addition, editing DSML models is usually performed using a dedicated environment providing services like model persistence, conformance checking, and others more advanced. We propose the use of patterns to address all these aspects, in order to facilitate and speed up their definition. By lack of space, we focus on patterns dealing with the abstract and concrete syntax, as well as the generation of modelling environments from them.

To deal with the abstract syntax, we propose *domain patterns*, gathering typical requirements of similar languages within a domain, and documenting their variability. Here, there may be patterns for workflow languages, arithmetical or logical expressions, variants of state machines, query languages, and component-based architectural languages, among others. A DSML may use several domain patterns, customized for a given need, and extended with other domain-specific concepts. These patterns may help to build a DSML more quickly and trustworthily, in a constructive way. As an example, Fig. 1 (left) shows a simplified domain pattern for state machines. Pattern elements have a cardinality, which governs how many times they can be instantiated (1 if no cardinality is specified). For instance, any application of the state-machine pattern should have one SimpleState, while it may lack InitialState and FinalState.

On the other side, *concrete syntax patterns* characterize families of similar representations [2], like textual, graphical, tabular or form-based. In the case of a graphical syntax, aspects like layouting or zooming may be configured. Moreover, concrete syntax patterns can be used to automate the generation of editors supporting the defined syntax (which otherwise should be implemented by hand), and can be attached to domain patterns in order to define different default visualization options for them. As an example, Fig. 1 (right) shows a simplified pattern for graph-based representation. This pattern permits assigning graphical elements (Circles, Rectangles, etc.) to elements in the DSML meta-model.

Altogether, in order to define DSMLs, we propose a reutilization-based, pattern-centric approach, which we have implemented in our prototype tool DSL-tao (http://jdelara.github.io/DSL-tao/). DSL-tao enables the construction of meta-models, where some meta-model parts can be defined through the application of existing patterns in a repository. Basic pattern application is performed in three steps. First, a pattern is selected and a wizard guides the designer in its application (see window 1 in Fig. 2 for the wizard of the state-machine pattern). In this step, variants and attached patterns can also be selected (see next section). Then, the designer can bind meta-model elements to pattern elements. Finally, the unbound pattern elements are automatically created new in the meta-model, annotated with their participant role in the pattern (window 2 in Fig. 2).

The next section presents two ways to describe and generate graphical environments for DSMLs using patterns.

3   Defining Graphical DSMLs Through Patterns

We propose two ways to describe the graphical syntax of a DSML. In the first one, domain patterns have attached a default visualization, which the DSML designer just reuses. This option profits from commonly agreed means to represent domain patterns (e.g., state machines, or component-based systems). In the second option, a dedicated wizard is used to apply a graphical pattern over the elements of the DSML meta-model. This approach is to be used when the DSML needs a non-predefined, or special syntax. Next, we present these two possibilities, as well as the environments generated through their use.

3.1   Using the Visual Syntax Attached to Domain Patterns

Domain patterns may have attached concrete syntaxes, accounting for typical representations of the domain concepts. For instance, Fig. 2 shows the application of the state-machine pattern. The pattern has three concrete syntax patterns attached: one for the standard graph-based representation, another for its representation as tables, and another using forms. Designers can select one of them. In this way, when the domain pattern is applied, the concrete syntax pattern will be automatically instantiated as well. Thus, this approach permits predefining a set of concrete visualizations, which can be reused "as is" by DSML designers.

3.2   Using the Dedicated Custom Wizard

Sometimes, the designer requires a fine grained control of the concrete syntax for the DSML, or he has not used domain patterns with attached concrete syntax. In such cases, the designer can still use a concrete syntax pattern to automate the generation of a modelling environment, for which he needs to map meta-model elements to concrete representations in the selected concrete syntax pattern. Since the application of concrete syntax patterns has many specificities

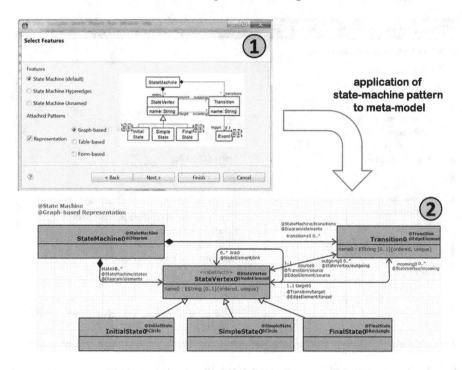

Fig. 2. The wizard for pattern application (1). Applied pattern (2).

(like selecting figures for nodes and decorations for edges), patterns may provide
dedicated wizards for their application. For instance, the graph-based concrete
syntax pattern has a customized wizard that implements heuristics to decide
which classes will be represented as nodes, which ones as edges, the attributes
to display, and the nodes that are containers of other nodes. Then, the designer
can refine the inferred concrete syntax and fine-tune the visual representation
for nodes and edges.

Figure 3 shows the wizard to customize the following heuristics:

- *Root strategies:* These are alternatives to select the root class to be used in
  diagrams. The root class is usually a class that contains all elements of the
  model, directly or indirectly. The strategy *Contains more classes* counts how
  many classes contain each class, and selects the one that contains more. The
  strategy *Class with no parents* suggests classes that are not contained in other
  classes. Both strategies are based on the tree of containment references defined
  in the meta-model. The last strategy (*Modularity pattern*) selects as root the
  meta-model classes annotated as Unit by a modularity pattern [5] (not shown
  in this paper) that allows organizing models in a modular way.
- *Label selection:* These heuristics are used to decide the data that node-like
  classes will display close to the node representation. The strategy *First string
  attribute* displays the first string attribute of the class, and *Identifier of the*

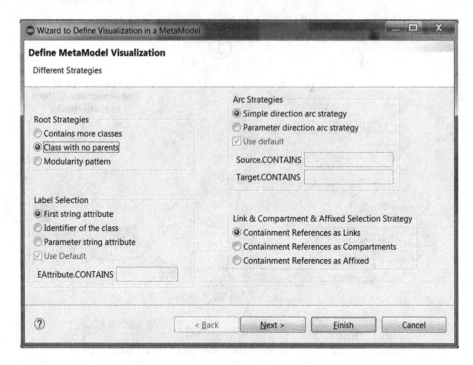

Fig. 3. Dedicated wizard for assigning a graph-based concrete syntax. Step 1: selection of heuristics.

class its identifier. The strategy *Parameter string attribute* receives several input strings, and selects the attribute whose name contains some of them.

– *Arc strategies:* They are used to select edge-like classes. In this case, we select the classes that define two non-containment references with lower bound 0 or 1, and upper bound 1. These two references will be mapped to the source and target of the edge representation for the class. While the first strategy (*Simple direction arc strategy*) selects the source and target references randomly, the second one (*Parameter direction arc strategy*) takes into account possible naming conventions (e.g., source or src for the source reference).

– *Link & compartment & affixed selection strategies:* These strategies identify the references that will be displayed graphically as edges, compartments or affixes. If the strategy *Containment references as links* is selected, all containment references will be represented as links. If the selected strategy is *Containment references as compartments*, they will be displayed as containers for the objects conformant to the type of the reference. Finally, if the chosen strategy is *Containment references as affixed*, the nodes will be placed on the border of another element.

The wizard uses the heuristics to infer the optimal concrete representation of meta-model elements, which are proposed to the designer in a second step (see Fig. 4). The, the designer is allowed to modify the inferred syntax, as well as

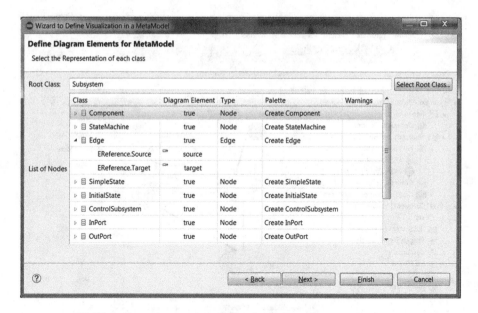

Fig. 4. Dedicated wizard for assigning a graph-based concrete syntax. Step 2: customization of inferred concrete syntax.

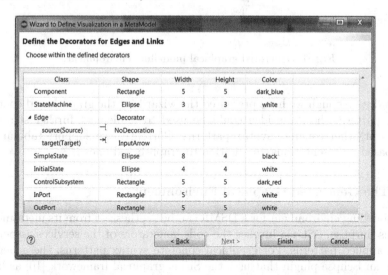

Fig. 5. Dedicated wizard for assigning a graph-based concrete syntax. Step 3: customization of appearance of nodes and edges (Color figure online).

fine-tune the concrete visualization for nodes and edges to customize the decorations for the start and end of edges, the types of figures for nodes, their size and colour. This last step is shown in Fig. 5.

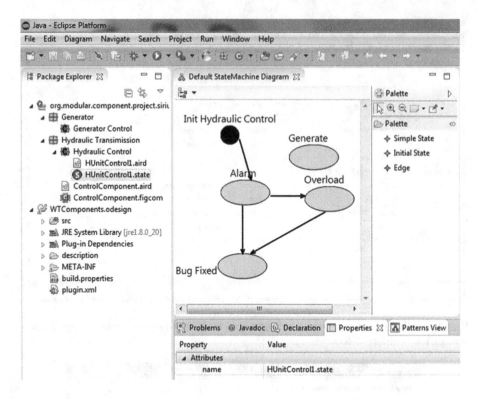

Fig. 6. Generated graphical modelling environment.

Finally, although we have presented the wizard for the graph-based concrete syntax pattern, the same idea could be used to implement further strategies for this or other concrete syntax patterns. Currently, we support tabular and form-based representations, in addition to graph-based ones.

3.3   The Generated Graphical Environment

The modelling environment for a DSML can be synthesized from its meta-model. For this purpose, DSL-tao invokes the code generators of the services associated to the applied patterns. For graphical concrete syntax patterns, the generator creates an Eclipse plugin that uses the Sirius graphical framework [10] as back-end. Thus, once the meta-model is annotated with the concrete syntax pattern, a Sirius *.odesign* model is generated. This model describes the shapes for nodes, the style for edges, the mappings of graphical elements to meta-model elements, the elements in the palette, and the actions to be performed when palette elements are invoked. Technically, this model is created using a model transformation. Then, the Sirius model is packaged in a plugin, which is contributed to the modelling environment of the DSML.

Figure 6 shows the generated graphical environment for the meta-model shown at the bottom of Fig. 2, which was created by instantiating the default concrete syntax pattern attached to the domain pattern for state machines.

4   Related Work

There are many tools to develop graphical modelling environments for different applications, like meta-CASE tools [8], diagram sketching [3] or multi-formalism modelling and simulation [4]. The advent of Eclipse has promoted frameworks to construct visual editors as plugins, like Tiger [1], GMF [6], Eugenia [9], Spray [11], Graphiti [7], or Sirius [10]. All these tools are model-based, except Graphiti which provides a Java API for coding. Some generate artefacts for other lower-level approaches, like Eugenia which is built atop GMF, and Spray atop Graphiti. In our case, DSL-tao produces graphical editors based on Sirius. All frameworks use code generation except Sirius, which is interpreted. The way of specifying the concrete syntax varies: Eugenia requires annotating the meta-model elements, Spray uses a textual DSL, GMF and Sirius require building models that describe the concrete syntax, and Graphiti requires programming. Our approach is closer to Eugenia, as our pattern applications result in meta-model annotations. However, our domain patterns can attach concrete syntax styles, which speeds up the generation of graphical environments. This feature is unique among the mentioned tools.

5   Conclusions and Future Work

We have presented a pattern-based approach to the development of graphical DSMLs. The approach is supported by a tool that permits applying patterns from a repository and the automatic generation of a modelling environment. We are currently working on defining new patterns, and developing further services for graphical environments like support for layers and abstractions.

Acknowledgements. Work supported by the Spanish Ministry of Economy and Competitivity (TIN2011-24139 and TIN2014-52129-R), the R&D programme of the Madrid Region (S2013/ICE-3006), and the EU commission (FP7-ICT-2013-10, #611125).

References

1. Biermann, E., Ehrig, K., Ermel, C., Taentzer, G.: Generating eclipse editor plug-ins using tiger. In: Schürr, A., Nagl, M., Zündorf, A. (eds.) AGTIVE 2007. LNCS, vol. 5088, pp. 583–584. Springer, Heidelberg (2008)
2. Bottoni, P., Grau, A.: A suite of metamodels as a basis for a classification of visual languages. In: VL/HCC, pp. 83–90 (2004)
3. Brieler, F., Minas, M.: A model-based recognition engine for sketched diagrams. J. Vis. Lang. Comput. **21**(2), 81–97 (2010)

4. de Lara, J., Vangheluwe, H.: AToM³: a tool for multi-formalism and meta-modelling. In: Kutsche, R.-D., Weber, H. (eds.) FASE 2002. LNCS, vol. 2306, pp. 174–188. Springer, Heidelberg (2002)

5. Garmendia, A., Guerra, E., Kolovos, D.S., de Lara, J.: EMF splitter: a structured approach to EMF modularity. In: XM@MoDELS, vol. 1239 of CEUR, pp. 22–31 (2014). CEUR-WS.org

6. GMF. https://wiki.eclipse.org/Graphical_Modeling_Framework

7. Graphiti. http://eclipse.org/graphiti/

8. Kelly, S., Tolvanen, J.: Domain-Specific Modeling - Enabling Full Code Generation. Wiley, Hoboken (2008)

9. Kolovos, D.S., Rose, L.M., Abid, S.B., Paige, R.F., Polack, F.A.C., Botterweck, G.: Taming EMF and GMF using model transformation. In: Petriu, D.C., Rouquette, N., Haugen, Ø. (eds.) MODELS 2010, Part I. LNCS, vol. 6394, pp. 211–225. Springer, Heidelberg (2010)

10. Sirius. https://eclipse.org/sirius/

11. Spray. https://code.google.com/a/eclipselabs.org/p/spray/

12. Xtext. http://www.eclipse.org/Xtext/

Computer-Computer Languages

Chapter 4 Computer Languages

Tree String Path Subsequences Automaton and Its Use for Indexing XML Documents

Eliška Šestáková[(✉)] and Jan Janoušek

Department of Theoretical Computer Science, Faculty of Information Technology,
Czech Technical University in Prague, Thákurova 9, 160 00 Prague 6, Czech Republic
{Eliska.Sestakova,Jan.Janousek}@fit.cvut.cz

Abstract. The theory of indexing texts is well-researched, which does not hold for indexing other data structures, such as trees for example. In this paper a simple method of indexing a tree for subsequences of string paths in the tree by finite automaton is presented. The use of the index is shown on indexing XML documents for XPath descendant-or-self axis inspired queries. Given a subject tree T with n nodes, the tree is preprocessed and an index, which is a directed acyclic subsequence graph for a set of strings, is constructed. The searching phase uses the index, reads an input string path subsequence Q inspired by the specific XPath query of size m and computes the list of positions of all occurrences of Q in the tree T. The searching is performed in time $\mathcal{O}(m)$ and does not depend on n. Although the number of distinct valid queries is $\mathcal{O}(2^n)$, the size of the index is $\mathcal{O}(h^k)$, where h is the height of the tree T and k is the number of its leaves. Moreover, we discuss that in the case of indexing a common XML document the size of the index is even smaller $\mathcal{O}(h \cdot 2^k)$.

1 Introduction

Indexing a data subject preprocesses the subject and constructs an index that allows to efficiently answer queries related to the contents of the subject. For example, occurrences of input patterns in the subject can be located repeatedly and quickly, in time typically not depending on the size of the subject.

The theory of text indexing, which is a result of Stringology research [5,7], is well-researched and uses many sophisticated data structures: suffix tree and suffix array are most widely used structures for text indexing for substrings, providing efficient solutions for a wide range of applications. The Directed Acyclic Word Graph [2], also known as suffix (or factor) automaton, is another elegant structure. An index of a text for subsequences is represented by a subsequence automaton [1], which is also referred as Directed Acyclic Subsequence Graph.

Indexing other data structures, such as trees for example, have not been developed in so many details and for so many indexing problems as in the case

J. Janoušek—This research has been partially supported by the Czech Science Foundation (GAČR) as project No. GA-13-03253S and by Technology Agency of the Czech Republic (TAČR) as project No. TA03010964 in α programme.

© Springer International Publishing Switzerland 2015
J.-L. Sierra-Rodríguez et al. (Eds.): SLATE 2015, CCIS 563, pp. 171–181, 2015.
DOI: 10.1007/978-3-319-27653-3_17

of indexing texts, although many practical applications serving as indexes of trees exist. Among others, a theory of indexing a data structure allows to understand the problem better, to find efficient solutions for particular indexing problems and to combine various indexes for the construction of indexes for unions, intersections, concatenations and other operations. In this last aspect, especially the use of the theory of formal languages and automata is very helpful.

An XML document represents a tree hierarchical structure. To be able to retrieve the data from XML documents efficiently, various query languages, such as XPath, XPointer and XLink, have been designed. Indexing the structure of XML data is an effective way to accelerate XML query processing. Therefore, several XML documents indexes have been introduced and we can divide them into the following four categories:

1. Graph-based methods construct a structural path summary that can be used to improve query efficiency, especially for single path queries. To this category we can classify following methods: DataGuides [9], 1-Index [17], PP-Index [20], F&B-Index [12] or MTree [18].
2. Sequence-based methods transform both the source data and query into sequences. Therefore, querying XML data is equivalent to finding subsequence matches. To this category we can classify following methods: ViST [22], PRIX [19].
3. Node coding methods apply certain coding strategy to design codes for each node in order that the relationship among nodes can be evaluated by computation. To this category we can classify, for example, XISS [13] method.
4. Adaptive methods can adapt their structure to suit the query workload. Therefore, adaptive methods index only the frequently used queries. To this category we can classify APEX Index [4], for example.

Each of these methods has its own advantages and disadvantages, however, shortcomings do exist: graph-based methods often possess a lack of supporting complex queries; sequence-based methods are likely to generate approximate solutions, thus requiring a great deal of validation; node coding methods are very difficult to be applied to ever changing data source and adaptive methods perform low efficiency on non-frequent queries.

In this paper we show that automata can be used efficiently for the purpose of indexing XML documents. Here, we consider to support only linear XPath queries. However, the techniques described here are relevant to the general XPath processing problem, for two reasons. First, processing linear expressions is a subproblem in processing more complex queries as we can decompose them into linear fragments. Second, this can be seen as a building block for more powerful processors, such as pushdown automata, that are able to process branching queries [14]. Moreover, it is easy to combine the index presented in this paper with our linear index of a tree for tree patterns, which represent connected subgraphs in a tree [11].

We introduce *Tree String Path Subsequences Automaton* that accepts and indexes all linear XPath queries using descendant-or-self axis (//), i.e., all subsequences of string paths in the tree. The searching phase uses the index, reads an input string path subsequence $\mathcal{Q}$ inspired by the specific XPath query of size m and computes the list of positions of all occurrences of $\mathcal{Q}$ in the tree $\mathcal{T}$. The searching is performed in time $\mathcal{O}(m)$ and does not depend on n. Although the number of distinct queries is $\mathcal{O}(2^n)$, the size of the index is $\mathcal{O}(h^k)$, where h is the height of the tree $\mathcal{T}$ and k in the number of its leaves. Moreover, we discuss that in the case of indexing a common XML document the size of the index is even smaller $\mathcal{O}(h \cdot 2^k)$.

2   Tree String Path Subsequences Automaton

We model an XML document as an ordered labelled tree where nodes correspond to elements, and edges represent element inclusion relationships. Hence, we only consider the structure of XML documents, and, therefore, will ignore attributes and the text in leaves.

A node in an XML tree model is represented by a pair $(label, id)$, where $label$ and id represent tag name and identifier, respectively. Without loss of generality, we have chosen to use a preorder numbering scheme to uniquely assign an identifier to each of the tree nodes.

Example 1. Consider following XML document $\mathcal{D}$. Figure 1 shows its corresponding XML tree model $\mathcal{T}$.

```
<HOUSES>
    <HOUSE name="Stark">
        <LORD>Eddard Stark</LORD>
        <SIGIL>Direwolf</SIGIL>
        <SEAT>Winterfell</SEAT>
        <VASSALS>
            <HOUSE name="Karstark">
                <LORD>Rickard Karstark</LORD>
                <SEAT>Karhold</SEAT>
            </HOUSE>
        </VASSALS>
    </HOUSE>
    <HOUSE name="Targaryen">
        <LORD>Daenerys Targaryen</LORD>
        <SIGIL>Dragon</SIGIL>
    </HOUSE>
</HOUSES>
```

As stated in the Introduction the *Tree String Path Subsequences Automaton* (TSPSA) efficiently evaluates all linear XPath queries over an XML document, where just node name test and descendant-or-self axis (//) are used. To simplify description of this XPath fragment we denote such class of queries by $XP^{\{//,name\ test\}}$.

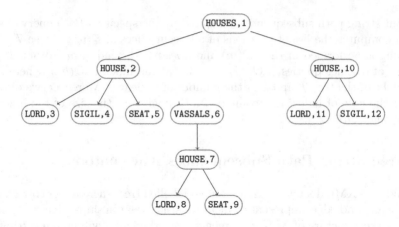

Fig. 1. XML tree model T of an XML document $\mathcal{D}$ from Example 1. Nodes in T are represented by pairs $(label, id)$, where $label$ and id represent element name and preorder identifier, respectively.

Definition 1 (XML Alphabet). *Let $\mathcal{D}$ be an XML document. An XML alphabet $\mathcal{A}$ of $\mathcal{D}$, denoted $\mathcal{A}(\mathcal{D})$, is an alphabet where each symbol represents a label of an XML element of $\mathcal{D}$.*

Example 2. Let $\mathcal{D}$ be the XML document from Example 1. The corresponding XML alphabet $\mathcal{A}$ of $\mathcal{D}$ is $\mathcal{A}(\mathcal{D}) = \{$HOUSES, HOUSE, LORD, SIGIL, SEAT, VASSALS$\}$.

Definition 2 (XPath Class of Queries). *Let $\mathcal{D}$ be an XML document. By $XP^{\{//,name\ test\}}$ we denote the language XPath class of queries over $\mathcal{D}$ generated by the following context–free grammar:*

$$G = (\{S\}, \mathcal{A}(\mathcal{D}) \cup \{//\}, \{S \rightarrow SS\} \cup \{S \rightarrow //a : a \in \mathcal{A}(\mathcal{D})\}, S)$$

Example 3. Let $\mathcal{D}$ be the XML document from Example 1 and $\mathcal{A}(\mathcal{D})$ be its XML alphabet as stated in Example 2. Consider following two XPath queries: $Q_1 = //$HOUSE$//$VASSALS$//$LORD, $Q_2 = /$HOUSES$//*$. The first query Q_1 belongs to $XP^{\{//,name\ test\}}$ class of queries over $\mathcal{D}$, whereas the second query Q_2 does not, since it contains syntax constructs out of the given class (i.e., / and *).

For an XML document of size n, the automaton processes a query of size m in time linear in m and not depending on n. The most similar approaches from XML indexing techniques are graph-based methods constructing a structural path summary [9, 20], which usually need further tree traversal to support queries containing // axis. Furthermore, our index is based on the idea of automata, which makes it well understandable.

Definition 3 (Subsequence of a String). *Let $X = x_1x_2 \ldots x_{|X|}$ be a string. A subsequence of X is any sequence of symbols x_i obtainable by deleting zero or more symbols from X.*

Example 4. Assume $X =$ HOUSES HOUSE LORD to be a string over an alphabet $\mathcal{A} = \{$HOUSES, HOUSE, LORD$\}$. There are 7 non-empty subsequences of the string X: HOUSES HOUSE LORD, HOUSES HOUSE, HOUSES LORD, HOUSE LORD, HOUSES, HOUSE, LORD.

As we attempt to index linear queries only, we can omit the branching structure and describe the XML tree model by means of its linear fragments, called string paths. To satisfy queries with // axis we are interested in (non-empty) subsequences of string paths.

Definition 4 (String Path). *Let T be an XML tree model with height h. A string path $P = n_1 n_2 \ldots n_{|P|}$, where $|P| \leq h$, of T is a linear path leading from the root $r = n_1$ to a leaf $n_{|P|}$. Each n_i of the path is associated with an identifier and label, denoted by $id(n_i)$ and $label(n_i)$, respectively. The identifier corresponds to a preorder number of the element.*

Definition 5 (String Path Alphabet). *Let P be a string path. A string path alphabet $\mathcal{A}$ of P, denoted $\mathcal{A}(P)$, is an alphabet where each symbol represents a label of a node in the path P.*

Definition 6 (String Paths Set). *Let $\mathcal{D}$ and T be an XML document and its XML tree model, respectively. A set of all string paths over T is called a string paths set, denoted by $\mathcal{P}_T = \{P_1, P_2 \ldots P_k\}$, where k is the number of leaves in the tree T.*

Example 5. Consider the XML tree model T in Fig. 1. We show its corresponding string paths set $\mathcal{P}_T$ below. Each node n_i in string paths is represented by its label (i.e., $label(n_i)$) and identifier (i.e., $id(n_i)$) in parenthesis.

$\mathcal{P}_T = \{$HOUSES(1) HOUSE(2) LORD(3),
        HOUSES(1) HOUSE(2) SIGIL(4),
        HOUSES(1) HOUSE(2) SEAT(5),
        HOUSES(1) HOUSE(2) VASSALS(6) HOUSE(7) LORD(8),
        HOUSES(1) HOUSE(2) VASSALS(6) HOUSE(7) SEAT(9),
        HOUSES(1) HOUSE(10) LORD(11),
        HOUSES(1) HOUSE(10) SIGIL(12)$\}$

Definition 7 (The Longest Query of a String Path). *Let $P = n_1 n_2 \ldots n_{|P|}$ be a string path and $X = //label(n_1)//label(n_2) \ldots //label(n_{|P|})$ be a string over the alphabet $\{//a : a \in \mathcal{A}(P)\}$ obtained from the path P. The string X is the longest $XP^{\{//,\text{name test}\}}$ query of the string path P, denoted by $maxq(P)$.*

Tree String Path Subsequences Automaton is a subsequence automaton [1] for a string paths set of an XML tree model representing the XML document being indexed. The automaton solving the problem of subsequences for both single and multiple strings is also referred as Directed Acyclic Subsequence Graph (DASG) and is further studied in [8,10]. Therefore, we propose an XML index problem to be another application area of DASG.

There are available three building algorithms for DASG for a set of strings: right-to-left [1], left-to-right [6] and on-line [10]. However, none of them is based on a subset construction, which gives a set of positions as answers of queries.

Therefore, we propose a construction of the Tree String Path Subsequences Automaton consisting of two steps. First, deterministic subsequence automata are constructed using subset construction, each accepting all non-empty subsequences of the longest query of a string path in $\mathcal{P}_T$. Second, the Tree String Path Subsequences Automaton is built using product construction.

To build a deterministic subsequence automaton, we propose two building algorithms: Algorithm 1 (construction using nondeterministic subsequence automaton with ε−transitions) and Algorithm 2 (direct subset construction of deterministic subsequence automaton). Resulting automata are used to build the Tree String Path Subsequences Automaton by Algorithm 3.

Data: A string path $P = n_1 n_2 \ldots n_{|P|}$.
Result: A deterministic finite automaton $M = (Q, A, \delta, 0, F)$ accepting all
      non-empty subsequences of a string $maxq(P)$.
1. Construct a deterministic finite automaton $M_1 = (Q_1, A, \delta_1, 0, F_1)$ accepting all prefixes of a string $maxq(P)$:
    (a) $Q_1 \leftarrow \{0, id(n_1), id(n_2), \ldots, id(n_{|P|})\}$,
    (b) $A = \{//a : a \in \mathcal{A}(P)\}$,
    (c) $\delta_1(0, //label(n_1)) \leftarrow id(n_1)$ and
        $\delta_1(id(n_i), //label(n_{i+1})) \leftarrow id(n_{i+1}), \forall i = 1, 2, \ldots, |P| - 1$,
    (d) $F_1 \leftarrow \{id(n_1), id(n_2), \ldots, id(n_{|P|})\}$.
2. Insert ε-transitions into the automaton M_1 leading from each state to its next state. Resulting automaton $M_2 = (Q_2, A, \delta_2, 0, F_2)$, where
    (a) $Q_2 \leftarrow Q_1, F_2 \leftarrow F_1$,
    (b) $\delta_2 \leftarrow \delta_1 \cup \delta'$ and $\delta'(0, \varepsilon) \leftarrow id(n_1)$,
        $\delta'(id(n_i), \varepsilon) \leftarrow id(n_{i+1}), \forall i \leftarrow 1, 2, \ldots, |P| - 1$.
3. Eliminate all ε-transitions. The resulting automaton is M_3.
4. Construct a deterministic finite automaton M equivalent to M_3 using standard determinisation algorithm based on a subset construction.

Algorithm 1. Construction of a deterministic subsequence automaton for a single string path using finite automaton with ε−transitions.

Example 6. Consider the XML tree model $\mathcal{T}$ illustrated in Fig. 1 and one of its string paths $P =$ HOUSES(1) HOUSE(2) SIGIL(4). Transition diagram of the subsequence automaton with ε−transitions for P constructed by Algorithm 1 is shown in Fig. 2. The resulting deterministic automaton is illustrated in Fig. 3.

We now introduce definitions of a set of occurrences of an element in a string path and function $ButFirst()$, which we need to refer to in the algorithm for direct subset construction of deterministic subsequence automaton presented afterwards.

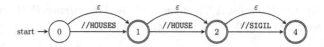

Fig. 2. Subsequence automaton with ε-transitions for a string path $P =$ HOUSES(1) HOUSE(2) SIGIL(4) from Example 6

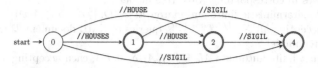

Fig. 3. Deterministic subsequence automaton for a string path $P =$ HOUSES(1) HOUSE(2) SIGIL(4) from Example 6

Definition 8 (Set of Occurrences of An Element in a String Path).
Let $P = n_1 n_2 \ldots n_{|P|}$ be a string path and e be an element node with label occurring at several positions in P (i.e., $label(n_i) = label(e)$ for some i). A set of occurrences of the element e in P is a totally ordered set $O_P(e) = \{o \mid o = id(n_i) \wedge label(n_i) = label(e),\ i = 1, 2, \ldots, |P|\}$. The ordering is equal to ordering of element prefix identifiers as natural numbers.

Definition 9 (ButFirst). *Let P and $O_P(e) = \{o_1, o_2, \ldots, o_{|O_P(e)|}\}$ be a string path and a set of occurrences of an element e in the string path P, respectively. Then we define a function $ButFirst(O_P(e)) = \{o_2, \ldots, o_{|O_P(e)|}\}$.*

Data: A string path $P = n_1 n_2 \ldots n_{|P|}$.
Result: A deterministic finite automaton $M = (Q, A, \delta, q_0, F)$ accepting all
        non-empty subsequences of a string $maxq(P)$.
1. $\forall n_i \in P$ compute $O_P(n_i)$.
2. Build finite automaton $M = (Q, A, \delta, q_0, F)$ accepting all prefixes of $maxq(P)$:
   (a) $Q \leftarrow \{q_0, q_1, \ldots, q_{|P|}\}$, $A \leftarrow \{//a : a \in \mathcal{A}(P)\}$, $F \leftarrow \{q_1, q_2, \ldots, q_{|P|}\}$,
   (b) $q_0 \leftarrow \{0\}$ and
       $\forall n_i$, where $i \leftarrow 1, 2, \ldots, |P|$:
       i. set state $q_i \leftarrow O_P(n_i)$,
       ii. add $\delta(q_{i-1}, //label(n_i)) \leftarrow q_i$,
       iii. $O_P(n_i) \leftarrow ButFirst(O_P(n_i))$.
3. Insert additional transitions into the automaton M:
   (a) $\forall i \in \{0, 1, \ldots, |P| - 1\}$ $\forall l \in \mathcal{A}(P)$:
       i. add $\delta(q_i, //l) \leftarrow q_s$, if there exists such $s > i$ where
          $\delta(q_{s-1}, //l) = q_s \wedge \neg\exists r < s : \delta(q_{r-1}, //l) = q_r$
       ii. $\delta(q_i, //l) \leftarrow \emptyset$ otherwise.

Algorithm 2. A direct subset construction of a subsequence automaton for a single string path.

Definition 10 *Let $\mathcal{D}$ be an XML document. A Tree String Path Subsequence Automaton accepts all $XP^{\{//,name\ test\}}$ queries of $\mathcal{D}$.*

Data: A string paths set $\mathcal{P}_T = \{P_1, P_2, \ldots, P_k\}$, where k is the number of leaves in corresponding XML tree model $\mathcal{T}$.

Result: A deterministic finite automaton $M = (Q, \{//a : a \in \mathcal{A}(\mathcal{D})\}, \delta, 0, F)$ accepting all $XP^{\{//,name\ test\}}$ queries of XML document $\mathcal{D}$ with XML tree model $\mathcal{T}$.

1. Construct finite automata $M_i = (Q_i, A_i, \delta_i, 0, F_i)$, each accepting a set of non-empty subsequences of a string $maxq(P_i)$ using Algorithm 2.
2. Construct deterministic Tree String Path Subsequences Automaton $M = (Q, \{//a : a \in \mathcal{A}(\mathcal{D})\}, \delta, 0, F)$ accepting a set of non-empty subsequences of each string $maxq(P_i)$ using product construction.

Algorithm 3. Construction of a Tree String Path Subsequence Automaton for an XML document $\mathcal{D}$ and its corresponding XML tree model $\mathcal{T}$.

Example 7. Let $\mathcal{D}$ and $\mathcal{T}$ be an XML document and XML tree model from Example 1 and Fig. 1, respectively. The corresponding Tree String Path Subsequences Automaton accepting all $XP^{\{//,name\ test\}}$ queries of $\mathcal{D}$, constructed by Algorithm 3, is shown in Fig. 4.

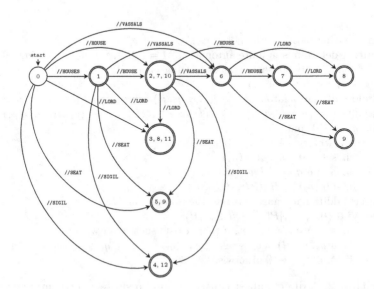

Fig. 4. Deterministic tree string path subsequences automaton

3   Evaluation of an Input Query

This section describes searching phase using the index. To compute positions of all occurrences of an input query Q in the XML tree model T of given XML document D, we simply run the Tree String Path Subsequences Automaton on the input query. Eventually, the answer for the input query is given by the subset contained in the terminal state of the automaton. If there is no transition that matches the input symbol, the automaton stops and rejects the input. Therefore, there are no elements in the XML document satisfying the query.

Example 8. Consider the XML document D and XML tree model T from Example 1 and Fig. 1, respectively. Suppose we want to evaluate following XPath query $Q_1 = $ //HOUSE//SEAT using Tree String Path Subsequences Automaton constructed by Algorithm 3. The transition diagram of the automaton for the document D is illustrated in Fig. 4.

Starting in the initial state, the automaton follows the transition for the first symbol of the input (i.e., //HOUSE) and goes to the state $(2, 7, 10)$. Next, the automaton continues reading the second symbol (i.e., //SEAT) and goes from the state $(2, 7, 10)$ to the state $(5, 9)$. Since the whole input is read and the automaton is in a final state, it returns positions $5, 9$ as the answer for the input query Q_1.

4   Time and Space Complexities

TSPSA effectively supports the evaluation of all $XP^{\{//, name\ test\}}$ queries of an XML document D. The number of such queries is clearly exponential in the number of nodes of the XML tree model T of D. Each state of TSPSA corresponds to an answer of a single query or a collection of queries. Although the number of different queries accepted by TSPSA is exponential, usually a lot of the queries are equivalent (i.e., their result sets of elements are equal).

Therefore, the equivalence problem of queries is closely related to the problem of determination the number of states of TSPSA. That is, if we know the number of unique query answers, we can construct a deterministic automaton answering all queries using exactly this number of states. On the other hand, we can obviously use the TSPSA to decide equivalence of two queries and even determine equivalence classes. The containment and equivalence problems for a fragment of the XPath query language was studied in [15,16]. For $XP^{\{//, name\ test\}}$ a PTIME containment algorithm was provided by Buneman et al. in [3].

From another point of view, we can examine the number of states of TSPSA as the size of DASG for a set of strings. For k strings of length h, the number of states can be trivially bounded by $\mathcal{O}(h^k)$ (size of a product of k automata with $\mathcal{O}(h)$ states). The running time for a query of length m becomes $\mathcal{O}(m)$. The lower bound for $k > 2$ texts in not known, while Crochemore and Tronicek in [8] showed that $\Omega(h^2)$ states are required for $k = 2$ at the worst case. Considering an XML index problem, k is a number of leaves in an XML tree model and h is its tree height.

Even, for a common XML document, in which the same nodes can only appear at the same level of the document, the size of the index is even smaller $\mathcal{O}(h \cdot 2^k)$ (Proof. In [21]). This is the asymptotic upper bound and we note that the size of the index is much smaller for many XML documents according to our experimental results [21].

5   Conclusion and Future Work

A simple method of indexing a tree for subsequences of string paths in the tree by finite automaton called Tree String Path Subsequences Automaton has been presented. This automaton is suitable for indexing XML documents for XPath descendant-or-self axis inspired queries and for easily combining the automaton with other tree indexes based on the automata theory. Given a subject tree T with n nodes, the tree is preprocessed and an index, which is a directed acyclic subsequence graph for a set of strings, is constructed.

The searching phase uses the index, reads an input string path subsequence Q inspired by the specific XPath query of size m and computes the list of positions of all occurrences of Q in the tree T. The searching is performed in time $\mathcal{O}(m)$ and does not depend on n. Although the number of distinct queries is $\mathcal{O}(2^n)$, the size of the index is $\mathcal{O}(h^k)$, where h is the height of the tree T and k is the number of its leaves. Moreover, we discussed that in the case of indexing a common XML document the size of the index is even smaller $\mathcal{O}(h \cdot 2^k)$.

There is a number of topics for future work:

- developing incremental building algorithm for TSPSA to efficiently adapt its structure to ever changing XML data source,
- proposing an indexing method able to support multiple XML documents,
- studying simulation techniques of nondeterministic finite automata and developing efficient simulation of TSPSA or its implementation using dynamic programming,
- extending our method to support more complex queries (e.g., including attributes, wildcards, branching etc.),
- creating a version of our index for cases when an XML document is so large that the index cannot be placed in the internal computer memory and therefore also an efficient use of external memory is needed.

References

1. Baeza-Yates, R.A.: Searching subsequences. Theoret. Comput. Sci. **78**(2), 363–376 (1991)
2. Blumer, A., Blumer, J., Haussler, D., Ehrenfeucht, A., Chen, M.T., Seiferas, J.I.: The smallest automaton recognizing the subwords of a text. Theor. Comput. Sci. **40**, 31–55 (1985)
3. Buneman, P., Davidson, S.B., Fan, W., Hara, C., Tan, W.-C.: Reasoning about Keys for XML. In: Ghelli, G., Grahne, G. (eds.) DBPL 2001. LNCS, vol. 2397, pp. 133–148. Springer, Heidelberg (2002)

4. Chung, C.-W., Min, J.-K., Shim, K.: Apex: an adaptive path index for xml data. In: Proceedings of the 2002 ACM SIGMOD International Conference on Management of Data, SIGMOD 2002, pp. 121–132. ACM, New York (2002)
5. Crochemore, M., Hancart, C., Lecroq, T.: Algorithms on Strings. Cambridge University Press, Cambridge (2007)
6. Crochemore, M., Melichar, B., Tronicek, Z.: Directed acyclic subsequence graph–Overview. J. Discrete Algorithms 1(3–4), 255–280 (2003)
7. Crochemore, M., Rytter, W.: Text Algorithms. Oxford University Press, Oxford (1994)
8. Crochemore, M., Troníček, Z.: On the size of DASG for multiple texts. In: Laender, A.H.F., Oliveira, A.L. (eds.) SPIRE 2002. LNCS, vol. 2476, pp. 58–64. Springer, Heidelberg (2002)
9. Goldman, R., Widom, J.: Dataguides: enabling query formulation and optimization in semistructured databases (1997)
10. Hoshino, H., Shinohara, A., Takeda, M., Arikawa, S.: Online construction of subsequence automata for multiple texts. In: Seventh International Symposium on String Processing and Information Retrieval, SPIRE 2000. Proceedings, pp. 146–152 (2000)
11. Janoušek, J., Melichar, B., Polách, R., Poliak, M., Trávníček, J.: A full and linear index of a tree for tree patterns. In: Jürgensen, H., Karhumäki, J., Okhotin, A. (eds.) DCFS 2014. LNCS, vol. 8614, pp. 198–209. Springer, Heidelberg (2014)
12. Kaushik, R., Bohannon, P., Naughton, J.F., Korth, H.F.: Covering indexes for branching path queries. In: Proceedings of the 2002 ACM SIGMOD International Conference on Management of Data, SIGMOD 2002, pp. 133–144. ACM, New York (2002)
13. Li, Q., Moon, B.: Indexing and querying xml data for regular path expressions. In: Proceedings of the 27th International Conference on Very Large Data Bases, VLDB 2001, pp. 361–370. Morgan Kaufmann Publishers Inc., San Francisco (2001)
14. Melichar, B., Janoušek, J., Flouri, T.: Arbology: trees and pushdown automata. Kybernetika 48(3), 402–428 (2012)
15. Miklau, G., Suciu, D.: Containment and equivalence for an xpath fragment. In: Proceedings of the Twenty-first ACM SIGMOD-SIGACT-SIGART Symposium on Principles of Database Systems, PODS 2002, pp. 65–76. ACM, New York (2002)
16. Miklau, G., Suciu, D.: Containment and equivalence for a fragment of xpath. J. ACM 51(1), 2–45 (2004)
17. Milo, T.: Index structures for path expressions. In: Beeri, C., Bruneman, P. (eds.) ICDT 1999. LNCS, vol. 1540, pp. 277–295. Springer, Heidelberg (1998)
18. Mark Pettovello, P., Fotouhi, F.: Mtree: an xml xpath graph index. In: Proceedings of the 2006 ACM Symposium on Applied Computing, SAC 2006, pp. 474–481. ACM, New York (2006)
19. Rao, P., Moon, B.: Prix: indexing and querying xml using prufer sequences. In: 20th International Conference on Data Engineering, 2004. Proceedings, pp. 288–299, March 2004
20. Tang, N., Yu, J.X., Ozsu, M.T., Wong, K.-F.: Hierarchical indexing approach to support xpath queries. In: IEEE 24th International Conference on Data Engineering, ICDE 2008, pp. 1510–1512, April 2008
21. Šestáková, E.: Indexing XML documents. Master's thesis, Czech Technical University in Prague, Faculty of Information Technology, Prague (2015)
22. Wang, H., Park, S., Fan, W., Yu, P.S.: Vist: a dynamic index method for querying xml data by tree structures. In: Proceedings of the 2003 ACM SIGMOD International Conference on Management of Data, SIGMOD 2003, pp. 110–121. ACM, New York (2003)

A Structural Approach to Assess Graph-Based Exercises

Rúben Sousa[(✉)] and José Paulo Leal

CRACS and INESC-Porto LA, Faculty of Sciences,
University of Porto, Porto, Portugal
up201001961@fc.up.pt, zp@dcc.fc.up.pt

Abstract. This paper proposes a structure driven approach to assess graph-based exercises. Given two graphs, a solution and an attempt of a student, this approach computes a mapping between the node sets of both graphs that maximizes the student's grade, as well as a description of the differences between the two graph. The proposed algorithm uses heuristics to test the most promising mappings first and prune the remaining when it is sure that a better mapping cannot be computed.

The proposed algorithm is applicable to any type of document that can be parsed into its graph-inspired data model. This data model is able to accommodate diagram languages, such as UML or ER diagrams, for which this kind of assessment is typically used. However, the motivation for developing this algorithm is to combine it with other assessment models, such as the test case model used for programming language assessment.

The proposed algorithm was validated with thousands of graphs with different features produced by a synthetic data generator. Several experiments were designed to analyse the impact of different features such as graph size, and amount of difference between solution and attempt.

Keywords: Automatic assessment · Graph comparison · Graph-based exercises

1 Introduction

Graphs are mathematical structures that model relationships among objects. They can be used in a wide range of domains such as network topology, software architecture, digital circuit design, just to mention a few. Diagrams are an apt example of a document type with a graph-based representation that requires automatic assessment. However, graphs can be used for assessing exercises where the relationship among parts is important but not determinant. Finite Deterministic Automata (FDA) and even programming languages are examples of this kind of assessment.

The assessment of an FDA should be twofold [2], based on the recognized strings and on the structure of the state automaton. If an FDA recognizes all the strings it should, and only those, then it must be correct. Otherwise, examples

© Springer International Publishing Switzerland 2015
J.-L. Sierra-Rodríguez et al. (Eds.): SLATE 2015, CCIS 563, pp. 182–193, 2015.
DOI: 10.1007/978-3-319-27653-3_18

of strings that should be recognized but are not, and vice versa, can be automatically generated. However, these examples seldom contribute to overcome the error. An helpful feedback must pinpoint what is wrong. For instance, what nodes are missing or what transitions must be removed. This can be achieved using graph assessment since the structure of an FDA can be represented by a state automaton [2].

Programming language assessment would also benefit from a similar approach. The standard way of assessing a program [4] is to compile it and then execute it with a set of test cases. A program is considered correct if it compiles without errors and the output of each execution is what is expected. If it is incorrect then the most this approach can provide are examples of input that generates wrong output. It cannot pinpoint the errors in the code of the program. An attempt to make this kind of assessment should be based on the structure of the program, specifically on its abstract semantic graph.

The ultimate goal of the research presented in this paper is to define a general methodology for assessing graph-based exercises, applicable to a wide range of domains including FDAs, programming languages but also diagrams. The objective of this paper is to propose a graph assessment algorithm and to evaluate its efficiency for graphs with the size typically used in automated assessment.

The proposed assessment algorithm is based on the *graph structure*. This means that it actually computes the mapping between the node sets of both graphs that best preserves their edges. To avoid checking a large number of mappings it iterates over them testing the most promising first. The mappings are iterated in an order that allows the algorithm to prune the majority of them, when it is ensured that the remaining mappings cannot produce a better mapping. The iteration order is driven by the types and properties of nodes.

The remainder of this paper is organized as follows. Section 2 surveys the existing literature on assessment of graph-based exercises. Section 3 describes the proposed algorithm, including the definition of the data structures to represent graph based exercises and their assessment. This approach is validated in Sect. 4 using a graph generator to test the applicability of the proposed algorithm. Finally, Sect. 5 highlights both the main contributions of this paper and the work ahead to apply this form of assessment in different scenarios.

2   Related Work

This section surveys the existing literature on automatic diagram assessment. To the best of the authors' knowledge, no general algorithm for the assessment of graph-based documents has yet been proposed. The existing proposals are targeted solely to diagrams, and focus mostly their labels rather than their structure.

Most of the available automatic diagram assessment systems were designed for a specific diagram type. Examples of these single diagram types addressed by existing systems are deterministic finite automata (DFA) [2,6], UML class diagrams [1,7], UML use case diagrams [10], Entity-Relationship diagrams [3], among others.

All these systems determine a mapping between nodes of solution graph and nodes the student's answer. The easiest approach is to use a fix set of labels in both graphs. For instance, the exercise descriptions used in assessment system proposed by Soler [7] for UML class diagrams requires fixing the class names used by students. Finding a mapping between the node sets of both graph is thus straightforward. A variant of this approach is the assessment in stages that Ali et al. [1] proposed. This system will not advance to the next stage until the current one is completely correct, otherwise it reports feedback on what is wrong or missing. The considered stages are: structural analysis, verification process and a language checking. The first stage compares the number of nodes, attributes and operation, and their types. The second stage checks if connections have the correct source and target type. Knowing that the graph structure is correct (by the two stages above), the system checks if the labels in nodes and attributes are nouns and in the operations are verbs.

The automata-base graph analyser of Shukur and Mohamed [6] works in a way that is similar to that presented above. The system does two types of evaluation: static and dynamic. The static analysis is made by comparing the global number of states, the number of initial and final states and the number of connections. The dynamic analysis is made by testing two sets of strings. One of the sets is composed by strings that the model should accept. If any of it is rejected, the graph is not correct. The second set is composed by strings that should be rejected by the system. So, by opposition, if any string is accepted the graph is not correct.

Thomas, Smith and Waugh [8,9] propose a system with similarities with the approach presented in this paper. It is a generic system able to handle different diagram types. Elements can be represented as boxes or circles and each connections as lines. The system tries to match those elements from students' answer to the elements of the solution. For each pair of nodes and edges is computed a similarity measure and with that value the system can assume what is (or not) a valid match. If the similarity is high, the system assumes it as correct. On the other hand, if the similarity is low the system is marked as not correct. The assessment algorithm described in the next section proposes a way to find the best match without these assumptions.

3   Graph Assessment Algorithm

The objective of the algorithm described in this section is to assess an exercise represented as an extended graph, by comparing it with a standard solution, represented also as an extended graph. The assessment consists in determining a set of differences between both extended graphs. These differences can be summarized in a grade, a numerical value within a fixed range (e.g. 0 to 100). If the set of differences is empty then the attempt of the student reaches the maximum grade; otherwise each difference introduces a penalty according to its type. For instance, a missing node may have a higher impact on the grade than a missing edge. Wrong types, missing or wrong properties have also their own penalties, depending on the graph-based language being assessed.

The basic approach is to determine the mapping between the node sets of both extended graphs that maximizes the grade. This can be solved by a simple generate and test strategy. If one generates all the possible mappings between both extended graphs, for each mapping one can determine the differences between both graphs and compute a grade. After iterating through all possible mappings one can select the one that produces the highest grade.

A solution and an attempt with equal sizes, i.e. with an equal number of nodes, is a particular case. In general these graphs have different sizes since the student may have omitted nodes or edges, or introduced unneeded ones. In this case the approach is to reduce the number of nodes in one graph until both have the same size; edges connecting the removed nodes are also removed.

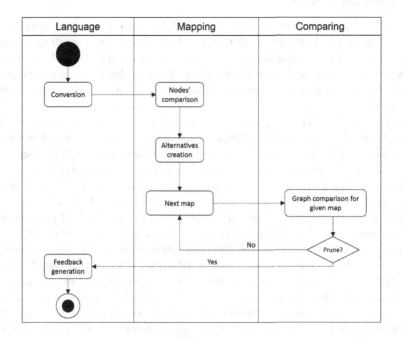

Fig. 1. Graph assessment activities

The graph assessment activities, depicted in Fig. 1, can be organized in three stages: language, mapping and comparing. The language stage depends on the actual graph language and needs to be configured for each case. The mapping stage produces all mappings connecting the nodes of both graphs. The graph comparison uses each mapping to computes a grade and a set of differences.

The process starts in the language stage when two files are converted into graphs, according to their type. Then all solution and attempt nodes are compared, and a list of ordered alternatives is created. This data is used for generating in a precise order the mappings in which graph comparison is based. The order in which mappings are generated is crucial for pruning the iteration over this collection of mappings. For each generated mapping a grade and a set of differences is computed. This process is repeated until the pruning condition is met. The computed grade and set of differences are further processed in the language layer in order to produce an adequate feedback for the graph language in question.

This section details several parts of the proposed algorithm. Subsection 3.1 introduces the definitions of extended graph and graphs differences, and defines the computation of a grade from a set of graph differences. Subsection 3.2 explains how node mappings are generated and pruned to enable an efficient assessment.

3.1   Data Structures

The proposed algorithm processes two *extended graphs*, a standard solution and a student attempt. A simple graph $G = (N, E)$ is defined by a set of nodes and a set of edges, where an edge is a pair of nodes. In an extended graph both nodes and edges have a *type* and a set of *properties*. An extended graph is a multigraph, in the sense that a pair of nodes may have more than a one edge, possible with different types.

Node and edge types capture the essential features of a graph-based language. Take UML diagrams for instance. Each kind of diagram combines nodes and edges of particular kinds. A use case diagram has as node types actor and use case, and as edge types associations, dependencies and generalizations. The features that are not captured by types are encoded as properties. Properties are simply name value-pairs. Consider an association in an UML class diagram; it may have a navigability, multiplicities, roles and other kinds of properties.

The assessment of an extended graph against another is a *set of differences*. The most relevant differences are detected when both graphs are made of the same size, such as insertion and deletion of nodes. The rest of the differences are computed based on a mapping between the nodes set of extended graphs with equal size. Consider a mapping $m : N \to N'$ and the nodes $a \in N$ from the extended graph used as solution. If a and $m(a)$ are indistinguishable then not difference is added to the set. Otherwise, a differences of a certain kind is signaled: if the types of a and $m(a)$ differ, $m(a)$ has a wrong type; if a property of a and $m(a)$ differs, a property insertion, deletion or wrong value is signaled.

The assessment restricted to nodes plays an important role in the proposed algorithm, since it is quicker to compute and is used in heuristics. However, a complete assessment must also consider edges. When nodes are removed to force both graphs to have the same size, the deletion of arcs connecting then is also marked. The rest of the arcs depends on the mapping. For each $(a, b) \in E$ is expected an $(m(a), m(b)) \in E'$ and vice-versa. Otherwise edge insertions, omissions, wrong type, as well as edge property differences, are also marked.

Given a set of differences it is possible to compute a grade. The empty difference set has the maximum grade (e.g. 100). Each kind of difference has a certain penalty and a grade is computed by subtracting these penalties to the maximum grade. Penalties depend on their kind and the size of the graph. In general a difference in a node should have a higher impact that a difference on a edge, but ultimately this depends on the graph-based language. There are a number of weights that have to be tuned for a particular language, based on actual grades given by expert teachers as benchmark. The same penalty has different impacts according to the solution graph size. For instance, a missing node will have higher impact on a small graph than on a large graph.

Grades computed from a set of differences are much more than just the final output of the assessment algorithm. They are essential to control it, in particular the node contribution to the grade, as is explained in the next subsection.

3.2   Node Mappings

The general strategy of assessing an extended graph against another is to determine a mapping between then that produces the higher grade. Due to the large number of possible mappings it is important to have heuristics to consider the most promising first and to have a criteria to prune most of them.

The node component of the assessment outweighs the edge component, although its computational complexity is much smaller. If both graphs have n nodes, there are n^2 pairs or nodes, although these can be combined in $n!$ mappings. If one iterates over the set or mappings by decreasing order of their node contributions to the grade (i.e. with increasing penalties), then the first mappings have higher chance of being the best than those appearing afterwards.

The first step for generating these mappings is to compute the contribution for the grade of individual node mappings. The initial mapping candidate is constructed from the individual mappings with best grade (less differences) for each node in the standard node set. In the example on Table 1 those individual mappings are represented in bold.

Table 1. Individual node comparison example

Attempt Solution	A'	B'	C'
A	**12**	8	11
B	8	**15**	7
C	10	7	**11**

The mapping candidate shown in Table 2 may actually be invalid if two different nodes are mapped in the same node. The rest of the individual mappings is generated by decreasing order of their contributions to grade.

It should be noted that the mappings are not created an then sorted, otherwise all the $n!$ mapping would have to generated. Instead, the successive mappings are generated by decreasing order or their contribution to the node grade from a list of node-to-node mapping. This list has only n^2 node-to-node mappings that are the building blocks of the mappings. This list is actually sorted in

Table 2. Best map found by comparing nodes on Table 1

$Match(A,A',12)$	$Match(B,B',15)$	$Match(C,C',11)$

decreasing order of their contribution to the node grade and it is used for finding replacements to the initial mapping.

New mappings are generated by replacing individual node mappings with an alternative. To ensure that the node contribution of the mapping decreases monotonously a sequence of target differences is explored in increasing order. The first grade difference to be explored is zero. That is, all sets of alternatives with a cumulative difference of zero to the best node grade are tested before all others. Then sets of alternatives with a cumulative difference of 1, 2, and so forth until all possible sets of alternatives are explored. The fact that grades are integer values is fundamental to this approach.

The number of mappings generated in this way is actually more than $n!$ (where n is the number of nodes of the graphs) since some of the mappings are invalid and are discarded. A mapping is invalid if it is not a bijective function. Hence, this process of generating mappings is only worthwhile if it can be pruned early and thus avoid generating most of the mappings.

After a number of iterations, the best mapping produces a grade g_{best}. The current mapping's grade is $g = g_{nodes} + g_{edges}$. If $g_{nodes} + g_{max\_edges} < g_{best}$ then it is sure that a better grade cannot be achieved with the remaining mappings since they all have a node contribution smaller than g_{nodes}.

Figure 2 shows the evolution of grades through successive iterations where g_{nodes} are represented in blue and g_{edges} in red. Solid rectangles represent components that contribute to the grade, while open rectangles represent the opposite. It should be noted that the total number of rectangles is constant, although the number of solid ones – the grade – oscillates; but the number of solid red rectangles – the g_{nodes} contribution to the grade – decreases monotonically. The grade computed in iteration 2 is better than the grade of iteration 1, due to an increased g_{edges} contribution, although with a smaller g_{nodes} contribution. This grade is not surpassed in iteration 3, but it cannot yet be declared the best. At iteration 4, even if contribution of g_{edges} reached its maximum, combined with the contribution of g_{nodes} it is less than the grade of iteration 2. Since the remaining mappings have a smaller or equal contribution of g_{nodes} to the grade, mapping generation can be pruned and the execution ended.

The node mappings generation process described above assumes that both graphs have equal sizes, which in general is not the case. If one graph has n nodes and the other has m nodes, with $n > m$, then there are $n!/m!$ different ways to make them equal. Again, the approach is to delete first the nodes that are least expected in the mapping, and pruning the tail of the node deletion list once it is certain that those alternatives cannot contribute to determine a mapping better than the one determined so far.

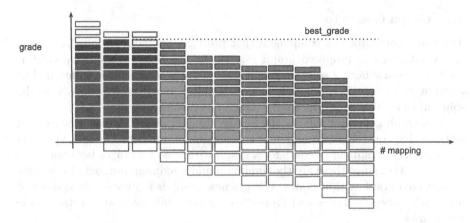

Fig. 2. Chart representing the evolution of grades (Color figure online)

The individual mappings are also used for selecting the order in which nodes are removed. For instance, if a single node has to be removed then the first attempt goes with the node that produces the worst contribution when mapped with any other, followed by the nodes with increasing contribution. Is a pair of nodes has to be removed then a similar approach is taken and is considered the combined contribution of these nodes. Since the groups of nodes to be removed are selected in the increasing order of their contribution to the grade, a similar pruning condition is used also in the graph reducing procedure.

4   Validation

The graph assessment algorithm described in the previous section was implemented in Java 1.8. This implementation was used in a number of experiments to validate the applicability of the proposed approach in the assessment of exercises on graph-based languages.

The validation was conducted using synthetic graphs. This approach contrasts with the validations described in the existing literature on diagram assessment systems, surveyed in Sect. 2. Most of the referred authors use actual exercises and student attempts, or a corpus with a large number of diagrams.

The reason for choosing synthetic data to validate this approach is twofold. Firstly, it is not intended for a specific graph-based language and should be adjustable to any graph-based languages that fits in the extended graph data model. Hence, its is important to test it with a wide range of settings, varying the number of types and properties, as will happen with different graph-based languages. Secondly, it is important to test the limits of the proposed approach, in terms of graph sizes and amount of difference between and attempt graphs, for which a large number of graphs pairs is required.

4.1   Graph Generator

The graph generator is a component that produces synthetic graphs for testing and validating the proposed graph assessment algorithm. This component is used to generate both a solution graph and attempts near a given solution. The attempt graph cannot be another random graph, it must be close enough to the solution to be able to produce a meaningful assessment.

The graph generator follows the builder design pattern. It has a number of settings that control of the minimum and maximum number of nodes, types and properties. The number of edges for a graph with n nodes ranges between $n - 1$ and $n(n + 1)/2$ since these are the minimum and maximum number of edges for a connected graph with n nodes. When a new graph is requested, its nodes and arcs with respective types and properties are randomly generated within these boundaries.

Graphs used in graph-based languages are typically connected graphs. Thus, the generator ensures that generated graphs are connected. It computes the connected components of the graph and, while it has more then one, it replaces a *redundant* edge in one component with an edge to a node in a different component. A redundant edge in a connected component is one that can be removed without breaking connectivity.

As explained above, the graph generator can also be used to produce graphs within the vicinity of a given graph, i.e. with a given number of variations, in number of nodes, edges, types and parameters. Within these boundaries the generator: inserts or removes nodes; changes types; inserts, removes or changes properties. Since the graphs produced this way are modelling student attempts, they may be disconnected graphs.

When producing a graph variant to model a student attempt, the generator produces also a set of differences. This set of differences uses the same type of data structure returned by the assessment method. Hence it is straightforward to compare the differences detected by the assessment method with those produced by the generator. This comparison validates the algorithm and its implementation.

Not all student attempts are wrong. Some may be equivalent to the standard solution, and this situation must also be tested. Nevertheless, it would be highly improbable for the two graphs to have nodes and arcs exactly in the same order. Comparing two exactly equal graphs could have an influence of the performance of the algorithm. Thus, the nodes and arcs of variant graphs are always shuffled, even if some differences were actually introduced.

4.2   Experiments

The implementation of the proposed graph assessment algorithm and synthetic graph generator described in the previous subsection were used in a series of experiments designed to answer the following questions.

Up to what graph size can this algorithm be used? The complexity of the graph homeomorphism problem is an NP problem neither known to be solvable in polynomial time nor to be NP-complete [5], but most likely the complexity is high enough to prevent the use of this approach on graphs above a certain size.

Do heuristics have a significant impact on performance? The heuristics were designed to explore the most promising mappings first, but they have an initial cost and depend of the effectiveness of the pruning criteria.

What is the impact of weights in performance? The algorithm is driven by grades and heuristics rely on the contribution of nodes to grades. The balance between the weight of node and edge grades is bound to influence performance.

What is the impact of domain specific data? The algorithm was designed to take advantage of the types and properties assigned to nodes and edges. This data makes node and edge easier to identify and the algorithm should perform better as more of it is available.

How dissimilar can solutions and attempts be? If the attempt and the solution are completely dissimilar then it makes no sense to compute differences between then and the grade should be zero. However, the assessment algorithm should perform well for attempts within a certain range of the solution.

The experiments that answered to these questions ran on a 4 cores computer with 8 i7-3630QM CPUs at 2.40 GHz, with 8 GByte of RAM. For each setting the experiment was repeated with 100 different random pairs of graphs. On most cases the assessment of a pair of graphs was executed well bellow 1/2 a second. Occasionally some pairs of graphs take a longer time hence a timeout of 2 s. In these case the assessment was considered *incomplete*, although the result obtained within the allotted time is correct in more than 50 % of the cases.

The first experiment addressed the size of the graphs that can be assessed with the proposed algorithm. Alur et al. argue that graphs in used exercises are usually smaller, with less than 10 nodes [2] and thus this complexity is not a serious problem. This appears to be the case in DFA, the domain they studied, and also many other domains, such as UML class and use case diagrams. However, an Entity-Relationship exercise to model a simple database may have more than 20 nodes, for instance. The results obtained with hundreds of equivalent graphs pairs show that the proposed algorithms deals with orders of up to 30.

Another experiment addressed the impact of pruning. For that purpose a variant of the mapping iterator was implemented. This iterator returns all the possibles mappings, without the initial overhead required by the iterator of the proposed algorithm. The rest of the algorithm was maintained unchanged. This algorithm was tested with equivalent graphs but could only complete the assessment of graphs of grade 6 or lower. Pairs of graphs with a larger number of nodes produce always an incomplete assessment. This compares with the use of the optimized iterator that can assess most graph pairs up to order 30.

The third experiment addressed the impact of weights, in particular the balance between the node and edge contribution to assessment. Since the heuristics rely on the node contributions to perform pruning, a larger contribution of edges decreases efficiency. It should be noted that the actual weights will depend of the specific graph-based language and on particular grading criteria defined by the teacher. In any event, it is expected that nodes contribute at least with half of the grade and in general with more than that. In fact, an equal weight of nodes and edges produces an assessment in less than 200 ms for graph pairs with up to 28 nodes, and the results improve as the weight of nodes is higher, as expected. The percentage of incomplete assessments is always less than 5 % and lowers as the weight of edges lowers.

The impact of domain specific data, i.e. the information provided by types and properties was also tested. Although the proposed algorithm is based on structure of the graphs, their nodes and edges, the heuristics use types and properties to distinguish them and improve efficiency. However, it should be noted that the number of types and properties depends on the graph-based language and cannot be controlled by the algorithm. As expected, the worst results were obtained with just one or two different types (a single type is equivalent to no types). With three to five types graph pairs of up to an order of 30 are assessed in 50 ms. The incomplete assessments reached 8 % for graphs with order 28 with 3 types, but was less than 2 % for all orders up to 30 with 4 or 5 types.

The experiments described above were performed with pairs of equivalent graphs to determine the impact of features. In these experiments it was checked that the algorithm found no differences between the graphs. The rest of the experiments were performed with different graphs and it was validated that the algorithm recovers the differences introduced by the generator. The algorithm was tested with solutions graphs with sizes up to 30 and attempt graphs with a size variation of up to 8 nodes. The execution time in these assessments is bellow 40 ms, with a tendency to increase with larger size differences. The number of incomplete assessments is bellow 5 % for solution graphs with up to 25 nodes and a different in number of nodes of less than 7.

5   Conclusions and Future Work

The main contribution of this paper is an algorithm for assessing graphs driven by their structure. It computes both a grade an explanation, a data object that can be serialized into a natural language text, or used as input for other systems. The assessment algorithm determines the best mapping between nodes in a solution graph and nodes in attempt graph. The mapping is the best in the sense that it maximizes the student's grade.

The algorithm validation ensured its efficiency for connected graphs with up to 30 nodes, which should cover the needs of exercise assessment. It suggests that automatic assessment systems for diagrams can be easily implemented based on this algorithm.

The next step is to validate assessment systems, rather than just the assessment algorithm, by using them with actual graph-based languages and actual students. An experiment is already scheduled for the last month of the current school year with students of a software architecture course. A parser of XML documents produced by the DIA diagram editor[1] is already in development.

The motivation for this assessment methodology is to blend it with other assessment methodologies, notably the test based assessment used with programming languages. This will require the study of the existing document type definitions for abstract semantic graphs of programming languages used in introductory programing courses such as Java, C/C++, Python and C#, and the existing tools for extracting abstract semantic graphs.

Acknowledgments. Project "NORTE-07-0124-FEDER-000059" is financed by the North Portugal Regional Operational Programme (ON.2 O Novo Norte), under the National Strategic Reference Framework (NSRF), through the European Regional Development Fund (ERDF), and by national funds, through the Portuguese funding agency, Fundao para a Cincia e a Tecnologia (FCT).

References

1. Ali, N.H., Shukur, Z., Idris, S.: A design of an assessment system for UML class diagram. In: International Conference on Computational Science and its Applications, 2007, ICCSA 2007, pp. 539–546. IEEE (2007)
2. Alur, R., D'Antoni, L., Gulwani, S., Kini, D., Viswanathan, M.: Automated grading of DFA constructions. In: Proceedings of the Twenty-Third International Joint Conference on Artificial Intelligence, pp. 1976–1982. AAAI Press (2013)
3. Batmaz, F., Hinde, C.J.: A diagram drawing tool for semi-automatic assessment of conceptual database diagrams (2006)
4. Douce, C., Livingstone, D., Orwell, J.: Automatic test-based assessment of programming: a review. J. Educ. Resour. Comput. (JERIC) **5**(3), 4 (2005)
5. Hell, P., Nesetril, J.: Graphs and Homomorphisms. Oxford University Press, Oxford (2004)
6. Shukur, Z., Mohamed, N.F.: The design of adat: a tool for assessing automata-based assignments. J. Comput. Sci. **4**(5), 415 (2008)
7. Soler, J., Boada, I., Prados, F., Poch, J., Fabregat, R.: A web-based e-learning tool for UML class diagrams. In: 2010 IEEE Education Engineering (EDUCON), pp. 973–979. IEEE (2010)
8. Thomas, P., Smith, N., Waugh, K.: Automatically assessing diagrams. In: Proceedings of the IADIS International Conference on e-Learning, vol. 2009 (2009)
9. Thomas, P., Waugh, K., Smith, N.: Automatically assessing free-form diagrams in e-assessment systems. In: 1st HEA Aiming for Excellence in STEM Learning and Teaching Annual Conference, Imperial College London (2012)
10. Vachharajani, V., Pareek, J.: A proposed architecture for automated assessment of use case diagrams. Int. J. Comput. Appl. **108**(4), 35–40 (2014). full text available

---

[1] http://dia-installer.de/.

Odin: A Service for Gamification of Learning Activities

José Carlos Paiva[1]([⊠]), José Paulo Leal[1], and Ricardo Queirós[2]

[1] Faculty of Sciences, CRACS & INESC-Porto LA,
University of Porto, Porto, Portugal
up201200272@alunos.dcc.fc.up.pt, zp@dcc.fc.up.pt
[2] CRACS & INESC-Porto LA & DI/ESEIG/IPP, Porto, Portugal
ricardoqueiros@eseig.ipp.pt

Abstract. Existing gamification services have features that preclude their use by e-learning tools. Odin is a gamification service that mimics the API of state-of-the-art services without these limitations. This paper describes Odin, its role in an e-learning system architecture requiring gamification, and details its implementation. The validation of Odin involved the creation of a small e-learning game, integrated in a Learning Management System (LMS) using the Learning Tools Interoperability (LTI) specification.

Keywords: Gamification · e-Learning · Game services · Interoperability

1 Introduction

The use of game concepts and mechanics in non-game contexts is an effective way to engage users. Gamification is currently a word of order in different domains, from marketing to e-learning [2]. The massive use of this approach led to the concept of gamification as a service, provided by major players such as Google[1] and Microsoft[2]. These services leverage on their large user base to provide support for game progress mechanics such as points, leaderboards and badges, without requiring a specific authentication from the client application.

Gamification services are a great advantage to small web and tablet based applications, in particular to games. The game progress mechanics features provided by these services are also relevant in e-learning. However, e-learning systems are typically deployed in environments with a single sign-on managed by an academic institution. It would be unacceptable to require students to have an account with a third party such as Google, for instance.

The purpose of the Odin service is to provide a gamification service similar to the state of the art, without requiring registration of the end users. Its API

---

[1] https://developers.google.com/games/services.
[2] http://azure.microsoft.com/en-us/documentation/services/mobile-services/.

© Springer International Publishing Switzerland 2015
J.-L. Sierra-Rodríguez et al. (Eds.): SLATE 2015, CCIS 563, pp. 194–204, 2015.
DOI: 10.1007/978-3-319-27653-3_19

is inspired in the Google Play Game Service (GPGS) with minor adjustments regarding user identification.

The remainder of this paper is organized as follows. Section 2 reviews the state of the art in game services. Section 3 introduces the Odin service, its design and implementation. Section 4 describes its evaluation using a small serious game as case study. Finally, Sect. 5 summarizes the contributions of this research.

2 Game Services

The video game industry is one of the fastest growing sectors in the worldwide economy [8]. According to the research company Gartner, global video game sales will reach \$111.1 billion in 2015, due in part to the growth in mobile game play and the recent release of the new generation of game consoles. In order to increase engagement and player retention, video games include several common features such as leaderboards and achievements. The massive use of this approach and the impressive growth of players led to the concept of gamification as a service, later materialized in Game Backend as a Service (GBaaS). The approach is simple. Instead of replicating the implementation of the game features in each version of the game for various platforms, GBaaS adhere to a service oriented architecture providing cross-platform game services that lets you easily integrate popular gaming features such as achievements, leaderboards, remote storage and real-time multiplayer in mobile games.

While the concept of "winners and losers" can hinder the motivation of students [7], gamification is currently being applied with relative success in e-learning [1,6]. The integration of game concepts in learning environments helps students to remain focused and to fulfill their course goals. However, the implementation of gamification in these domains is often trapped in ad-hoc solutions or supported by specific platforms (for instance, the badges in Moodle), instead of using approaches such as those provided by GBaaS.

In the following subsections we briefly summarize the main common game features that can be applied to the teaching-learning process. Then, we compare six GBaaS regarding social and technical features. This study is part of an effort to select an GBaaS on which to base the development of a service for gamification of learning activities.

2.1 Game Concepts

Games are more interesting when players are able to achieve goals and compete against other players. These features foster retention and competitiveness, and are applicable also in the gamification of e-learning activities. The following list enumerates the most common game concepts:

Leaderboards are databases that keep scores. They allow users to post their scores in a game and compare themselves with other players' scores. They measure the success of a player in a game.

Achievements are goals/challenges set in a game that players managed to accomplish. Achievements give players a motivation to keep playing, to earn as many as possible, and a way to compare themselves with other players. The fulfilment of a goal may enhance the status of the player or unlock access to other levels, for instance.

Multiplayer is a play mode that allows several players to simultaneously cooperate or compete in a game. This feature supports a range of other subfeatures, such as challenges, where players compete each other on either a score challenge or an achievement challenge, and matchmaking games in realtime, turn-based, or self-hosted matches.

Saved games allow the remote storage (in the cloud) of game data, for instance, the state and the players' progress in the game.

Quests are periodic game challenges that players can complete to earn rewards. This way, developers can launch periodic challenges to their gaming communities.

Gifts allow players to send/request game resources or items to/from friends (for instance, in their Google+ circles).

Matchmaking automatically sets up game matches and finds opponents based on parameters set by the game developer. Usually only a specific number of players can be matched at the same time.

2.2   Game Backend Services

A **Backend-as-a-service (BaaS)** is a cloud computing service model acting as a middleware component that allows developers to connect their Web and mobile applications to cloud services via application programming interfaces (API) and software developers' kits (SDK). BaaS features include cloud storage, push notifications, server code, user and file management, social networking integration, location services, and user management as well as many other backend services. These services have their own API, allowing them to be integrated into applications in fairly simple way [3].

A **Game-Backend-as-a-Service (GBaaS)** is a subset of a BaaS that includes cross-platform solutions for the typical game concepts identified in the previous subsection. During the development process of a game (or a generic application) developers must choose between building their own back-end services or using an available game back-end platform. This last option is usually preferred since GBaaS include several services specifically tailored for game development. These services allow developers to focus on the game logic by freeing them from implementing boiler plate features.

The following subsections compare several GBaaS according to their social and technical features. Given the number of GBaaS found (32) it would be impracticable to study them all. Therefore, six GBaaS were chosen: Google Play Game Services, Yahoo Bakend Game Service, GameUp, GameSparks, Fresvii and Photon. These features are summarized in Table 1.

Social Game Features. The studied GBaaS provide developers with social game services accessed through cross-platform API. These features make the gameplay more competitive and collaborative, and improve social engagement.

Table 1. Social and Technical game features

Types	Features	Google	Yahoo	GameUp	GSparks	Fresvii	Photon
Social	Leaderboards	yes	no	yes	yes	yes	yes
	Achievements	yes	yes	yes	yes	no	no
	Multiplayer	yes	yes	no	yes	yes	yes
	Save data	yes	yes	yes	yes	yes	yes
	Quests	yes	no	no	yes	no	yes
	Gifts	yes	yes	no	yes	no	yes
	Matchmaking	no	no	yes	no	yes	yes
Technical			Yahoo		Face		
	Auth	G+	Face	Face	Twitter	Face	Face
	WS	REST	-	REST	REST	-	REST
	Res. format	JSON	-	JSON	JSON	-	JSON
					ActionScript		
			ActionScript		C++		
			iOS		Cocos2D		
		Andriod	Andriod	Andriod	JavaScript	Andriod	Andriod
		iOS	C#	iOS	Marmalade	iOS	.NET
	Platforms	C++	Unity	Unity	Unity	Unity	Unity

Analysing Table 1 one concludes that almost all GBaaS supports leaderboards, multiplayer game mode and cloud storage. Other features such as quests and matchmaking are not yet widely supported, probably due to their novelty.

Technical Game Features. The studied GBaaS offer cloud services through API and SDK to various platforms. Regarding authentication almost all GBaaS use the same strategy. Before the game can make any calls to the game services, it must first establish an asynchronous connection with the backend servers and authenticate within the game services. Some GBaaS requires that the players have an account on specific backends (GPGS requires that users have a Google account). Others, such as GameSparks, provides a simple mechanism that allows games to implement social login without any additional code, allowing gamers, for instance, to sign in using a Facebook or Twitter account, and start playing.

The majority of the GBaaS provides a HTTP RESTful API. The format of the data in all HTTP store operations (PUT and POST) is required to be valid JSON. All response data from the GBaaS comes back also in JSON format. Regarding the REST API reference, the authors' opinion is that GPGS is the most complete and best documented API.

In complement to the REST API most GBaaS support also mobiles. There are examples of SDKs for Android, iOS, and even FirefoxOS (GameUp) mobile native apps. Game engines are also supported and most GBaaS offer SDKs for major game engines such as Unity, and also for cross-platform game development tools such as Marmalade and Cocos2D.

3   Odin

This section describes Odin, a gamification RESTful Web Service to be used by educational institutions. It provides (1) score submissions, (2) leaderboards listing, (3) quests for players, (4) awards to players for in-game accomplishments as well as some minor services to manage institutions, players, leaderboards, quests and achievements.

Odin is based on a standard gamification API but has a different approach regarding authentication. Institutions, rather than end-users, are the ones that require authentication. Once an institution is authenticated, Odin grants it permission to manage scores, quests and achievements in its users.

The next subsections present the architecture of Odin and its main components, and describe its data model and service API.

3.1   Architecture

Odin is a RESTful Web Service that allows institutions to consume gamification resources from their web applications. The web applications initialize sessions in Odin through authentication built on top of OAuth2 authorization protocol[3]. Then they request particular actions to the server identified by a specific URI and an HTTP method such as POST, GET, PUT or DELETE.

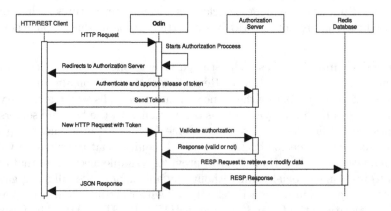

Fig. 1. Sequence diagram representing a common request to Odin

Figure 1 presents a sequence diagram that summarizes the interactions of Odin with other systems when a request is made by the client. Firstly, the

---

[3] http://oauth.net/2/.

HTTP request made by the client is subject to a security filter that checks if the institution is authenticated. If the institution is not authenticated or authorized to access Odin resources it is redirected to the authorization server where it will authenticate and approve the release of a token with the authorization proof. The generated token (with expiration time) is sent to the client and it (client) presents the access token to Odin.

When the client is authenticated and authorized, it is passed to the JAX-RS REST interface implemented using Jersey (described in the next subsection) and forwarded to the mapped resource. From the resource layer it is forwarded to the service layer, passing through a security layer which intercepts it to check authorization and roles, ensuring that only authorized institutions have access to the services.

The service layer responds to the request with the data persisted on Redis (described in the next subsection) through the Jedis client (using REdis Serialization Protocol) and Ohm library implementation for Java. The response sent to the client is a JSON object representing the resource type modified or requested by it (each resource type may have one or more data representations). Whenever a fresh token is needed, the client can request it from the Authorization Server.

3.2 Frameworks and Tools

Odin uses Jersey[4], an open-source framework that is the reference implementation of the Java API for RESTful Web Services, extending it with additional features and utilities to further simplify RESTful service. Among other features, Jersey provides a *Core Server* to build RESTful services based on annotations, support for JSON and to the Java Architecture for XML Binding, as well as a *Core Client* to easily create a client that can communicate with REST services.

Data storage relies upon Redis NoSQL database[5] that provides an open-source and advanced key-value storage and cache solution. It is an high performance alternative to the traditional Relational Database Management Systems (RDBMS) [5] to store and access large amount of data. Redis is sometimes described as a data structure server since keys can contain strings, hashes, lists, sets and sorted sets. As a NoSQL database it focus on performance and scalability rather than in guaranteeing the atomicity, consistency, isolation and durability (ACID) properties. Redis was selected for backend due to its hability to store large amounts of non critical data very efficiently.

In order to integrate Redis in Odin the data layer resorts to the Jedis client[6], as well as of an object-hash mapping library, named JOhm[7], to store and retrieve objects from Redis with an higher level of abstraction and thus simplicity. JOhm is the Java implementation of the well-known Ohm library[8] and aims to be

---

[4] https://jersey.java.net/.
[5] http://redis.io/.
[6] https://github.com/xetorthio/jedis.
[7] https://github.com/agrison/johm a fork from https://github.com/xetorthio/johm.
[8] http://ohm.keyvalue.org/.

minimally-invasive, relying only on reflection aided by annotation hooks for persistence.

3.3   Data Model

The data model of Odin consists of seven main entities: institution, player, leaderboard, score, quest, achievement and session, related as denoted in the UML class diagram of Fig. 2.

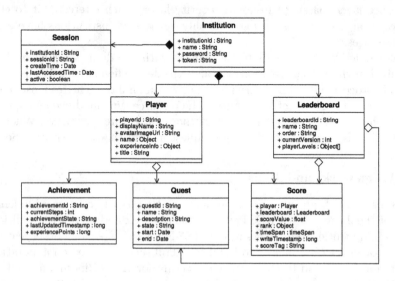

Fig. 2. Class diagram of the data model of Odin

An institution is the entity that manages games and all related data, and so it is the one which needs authentication and/or authorization. Thus, it needs to store an id and password to authenticate, and also a token to check the validity of the session. Whenever an institution authenticates a session is created and linked to it (through the *institutionId*). This session contains the creation time, last access time and a state indicator (active or inactive).

The institution needs to represent its students. As this is a gamification model they are abstracted to players, and so they will have a *playerId* that identifies them to the institution, a *displayName* that is the name to show on the leaderboard, a full name and a representation of his experience info with level, points acquired and needed points to level up.

As the player progresses in the game, (s)he will possibly win achievements. An achievement has a number of required steps and a state (hidden, revealed or unlocked). When a player reveals one, he receives the number of experience points associated.

A player can also accept and fulfill quests. A quest is characterized by a name, a description, a state (upcoming, open, accepted, completed, failed, expired or deleted) and a start and end date.

One of the most important parts of this model is the leaderboard. It contains more than a list of sorted scores, it contains data related to a game, such as a list of info on the levels available in the game/leaderboard. These parts are joined since it is required a single leaderboard to each game, and they depend on the existence of each other.

Scores related to a leaderboard and a player, are also stored. Each score has a floating point value, a timespan (daily, weekly or all time score), a timestamp and a rank (its position on the leaderboard).

3.4 Service API

The integration of Odin with other systems relies on REST calls to set and retrieve data. It follows the Google Web API Reference[9] for achievements, leaderboards, players, quests and scores resources. The only differences are that all these resources' URI paths are relative to gamify/institutions/**institutionId**. Also when an authenticated player is referenced in a function, it is replaced by a sub-path of the form /players/**playerId** right after **institutionId** in the resource path URI.

The institution resource is added to the set of resources. It contains the functions shown in Table 2.

Table 2. Institutions resource API reference. URIs are relative to/gamify

Function	HTTP request
insert	POST/institutions
get	GET/institutions/institutionId

The insert function inserts the institution given in the request body. The get function retrieves the institution resource given its id.

4 Evaluation

For validation of the gamification service described in the previous section, a simple multiplication game was created. This game – MathGamify – can be used by primary school children to learn multiplication tables. MathGamify generates two random numbers. The first number between 1 and the current game level and the second number between 1 and 10. Then the student/player has the opportunity to answer the multiplication value of the two numbers. The score is

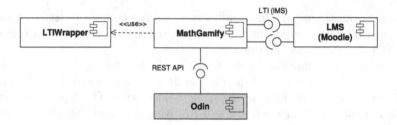

Fig. 3. MathGamify component diagram

accumulated in the ratio of the player's level until player misses, in which case the score is reset to zero.

Figure 3 presents the component diagram of MathGamify. MathGamify acts as a tool provider to a Learning Management System (LMS). The integration of MathGamify with the LMS relies on the Learning Tools Interoperability (LTI) specification. When the LMS launches MathGamify the LTI parameters are sent as part of the HTTP POST request. On request reception MathGamify uses the LTI Wrapper [4] package to process LTI communication and extract user id, name and level. The last is a custom parameter defined on the external tool configuration of the LMS.

MathGamify consumes two types of resources from Odin: score submission and listing of scores. Once the player answers a question, MathGamify communicates the score to Odin, using Jersey Client to issue the REST call, and the grade to the LMS using LTI. This grade is a value between 0 and 1, calculated by the following way: if there is a custom parameter custom_max_score then it is the score divided by custom_max_score, otherwise it is the number of correct answers divided by the total number of tries. When MathGamify initializes its GUI, and every time a score is submitted, the score listing is updated with the data returned from Odin.

One of the key components is the LTI Wrapper that implements both sides of the LTI communication. This component receives LTI requests from LMS and issues LTI requests to LMS.

The GUI component of MathGamify was developed using Google Web Toolkit (GWT), an open source Java software development framework that allows a fast development of AJAX applications in Java. The GWT code is organized in two main packages, the server and the client. The server package includes all the service implementations triggered by the user interface. These implementations are responsible of (1) the logic of the game, (2) communication with Odin and (3) communication with LMS through LTI wrapper. The selected LMS was Moodle 2.8[10].

The implementation of MathGamify demonstrates the efficacy of the proposed approach in coping with the extra requirements of a serious game

---

[9] https://developers.google.com/games/services/web/api/index.
[10] https://moodle.org/.

integrated in a typical e-learning ecosystem, where authentication is provided by an LMS. To complement its validation, Odin was also tested regarding its efficiency.

The latency of the Odin service was tested in two of its functions: (1) submit a single score and (2) list all scores in a leaderboard. Each test consisted of 1000 samples of calls to the same function, and all numbers stated below are averages per sample.

Initially the tests were run locally on the same machine as the Odin server, using Grizzly Test Container provided by Jersey, so it had no network latency. The average time to (2) was around 40 ms (leaderboard had 6 scores when the test was running). In the worst case it took 461 ms. The test (1) spent an average time of 22 ms and the worst case took 385 ms.

The same tests were repeated on an external server. During these tests an average network latency of 23 ms was observed. In this setting test (1) consumed an average time of 67 ms. The average time to (2) was 587 ms (the leaderboard had 1000 scores).

The tool used to measure time spent was ContiPerf[11], a lightweight testing utility that allows the user to easily turn JUnit 4 test cases to performance tests. It is based on annotations as the JUnit 4's test configuration.

5   Conclusions

Game concepts and mechanics are an useful way to engage students in e-learning activities. These kind of features are already provided by game backend services that can leverage on their authentication services and massive user base. However, gamification services that rely on external authentication are not adequate for e-learning systems that already operate on a single sign-on ecosystem.

Odin is a gamification service developed for requirements of e-learning systems. It was designed to authenticate clients rather than end-users and thus can be integrated with the e-learning systems typically found in educational institutions.

The MathGamify system is a proof of concept, that illustrates how serious games acting as tool providers for an LMS interact with the services of Odin. The authors plan to integrate Odin in an e-learning tool for formative assessment in online and hybrid courses. This tool will interface with Odin to support the creation of leaderboards, introduce timed challenges, reward students for their achievements, among others.

Odin itself will be subject to improvements. The current version provides web services for exposing the gamification service to clients. The next version will provide also a web interface to register institutions and allow them to manage their resources. A Reference API documentation for Odin will also be created.

Acknowledgments. Project "NORTE-07-0124-FEDER-000059" is financed by the North Portugal Regional Operational Programme (ON.2 O Novo Norte), under the

---

[11] http://databene.org/contiperf.

National Strategic Reference Framework (NSRF), through the European Regional Development Fund (ERDF), and by national funds, through the Portuguese funding agency, Fundação para a Ciência e a Tecnologia (FCT).

References

1. Burguillo, J.C.: Using game theory and competition-based learning to stimulate student motivation and performance. Comput. Educ. **55**(2), 566–575 (2010). http://dx.doi.org/10.1016/j.compedu.2010.02.018
2. Hamari, J., Koivisto, J., Sarsa, H.: Does gamification work? – a literature review of empirical studies on gamification. In: 2014 47th Hawaii International Conference on System Sciences (HICSS), pp. 3025–3034. IEEE (2014)
3. Janssen, C.: Backend-as-a-service (baas). Technical report, Techopedia (2014). http://www.techopedia.com/definition/29428/backend-as-a-service-baas
4. Queirós, R., Leal, J.P., Campos, J.: Sequencing educational resources with seqins. Comput. Sci. Inf. Syst. **11**(4), 1479–1497 (2014)
5. Seeger, M., Ultra-Large-Sites, S.: Key-Value Stores: A Practical Overview. Computer Science and Media, Stuttgart (2009)
6. Siddiqui, A., Khan, M., Akhtar, S.: Supply chain simulator: a scenario-based educational tool to enhance student learning. Comput. Educ. **51**(1), 252–261 (2008). http://dx.doi.org/10.1016/j.compedu.2007.05.008
7. Vansteenkiste, M., Deci, E.L.: Competitively contingent rewards and intrinsic motivation: can losers remain motivated? Motiv. Emot. **27**, 273–299 (2003). doi:10.1023/A:1026259005264. http://dx.doi.org/10.1023/A:1026259005264
8. Zackariasson, P., Wilson, T.: The Video Game Industry: Formation, Present State, and Future. Taylor & Francis, New York (2012). http://books.google.pt/books?id=lgiQNdc-DOwC

SplineAPI: A REST API for NLP Services

Nuno Vieira[1], Alberto Simões[1,2](✉), and Nuno Ramos Carvalho[1]

[1] Centro Algoritmi, Universidade do Minho, Braga, Portugal
nunovieira220@gmail.com, ambs@ilch.uminho.pt, narcarvalho@di.uminho.pt
[2] Centro de Estudos Humanísticos, Universidade do Minho, Braga, Portugal

Abstract. Modern applications often use *Natural Language Processing (NLP)* techniques and algorithms to provide sets of rich features. Researchers, who come up with these algorithms, often implement them for case studies, evaluation or as proof of concepts. These implementations are, in most cases, freely available for download and use.

Nevertheless, these implementations do not comprise final software packages, with extensive installation instructions and detailed usage guides. Most lack a proper installation mechanism and library dependency tracking. The programming interfaces are, usually, limited to their usage through command line, or with just a few programming languages support.

To overcome these shortcomings, this work aims to develop a new web platform to make available a set of common operations to third party applications that can be used to quickly access *NLP* based processes. Of course this platform still relies on the same tools mentioned before, as a base support to specific requests. Nevertheless, the end user will not need to install and learn their specific Application Programming Interfaces (API). For this to be possible, the architectural solution is to implement a *RESTful API* that hides all the tool details in a simple API that is common or, at least, coherent, across the different tools.

Keywords: Natural language processing · REST API · Web service · DSL

1 Introduction

Natural Language Processing (NLP) techniques are being used in very different types of applications.

Some companies are mining social communities to find out what their customers think about their products or services [3]. Others are making their information available in different languages by using machine translation techniques [9]. Newspapers and other news agencies, are using NLP techniques to summarise news and cluster them by specific areas, or based on their similarities [5].

Any one of these applications require a stack of NLP tools to work. This stack can be very different from tool to tool, but might include common tasks like: language identification, text segmentation, sentence tokenization, part of speech

© Springer International Publishing Switzerland 2015
J.-L. Sierra-Rodríguez et al. (Eds.): SLATE 2015, CCIS 563, pp. 205–215, 2015.
DOI: 10.1007/978-3-319-27653-3_20

tagging, dependency parsing, probabilistic translation, dictionaries querying, or named entity detection, just to mention some [6].

Although there are some NLP toolkits that include a good number of tools for most of these tasks [1,4], developers are likely to need other tools that are not directly available. This leads to the installation of different tools. If the developers need to support a wide range of languages, this list of tools is prone to grow, as some tools are not language independent or because they do not include training data for some of the required languages.

These requirements lead to the need of installing a variety of tools to have a complete NLP stack. Unfortunately, most of these installations are not as simple as they should be, as most of their developers are more interested in using the tools and adding new features than to document their usage and installation, or to provide good installation procedures. This leads to the need of dealing with different kinds of installation problems, and to learn each tool application programming interface (API).

Although our NLP team is small, we have been dealing with this problem for some time, and therefore, we are proposing a tool and a service to hide all these details from the end-user, making these libraries available as web services based in the REST philosophy. Of course that, if the web services are, themselves, using those same tools, someone will need to deal with the installation procedure, and will need to learn its usage. But if this process could be done only once, and the installed tools are available as a simple web service, application development is faster, and application deployment gets easier.

As a side benefit, having a different server running some tools, helps in distribution. Even if at the moment we have the system working on a single server, it is simple to distribute the tools between different machines.

Nevertheless, the process of making these tools available through a web service is not straightforward, as one needs to deal with timely processes, that can not be served easily using a single HTTP request, given timeouts; problems on service abuse; problems on load distribution, and others.

In this paper we present SplineAPI, that is both a service, that we are making available for free, and a platform, for anyone to replicate this kind of service in their own servers. Section 2 will compare our proposal with other services already available on the *Web*. Section 3 includes a presentation of our design goals as well as the SplineAPI architecture and implementation. Section 4 concludes with future work.

2   Related Work

The idea to make APIs available through web services is not new. There are several platforms that make NLP processes available online, each with its own characteristics and targeting different kinds of users. They range from simple tools that allow a single kind of task to be performed, to fully featured sites with a diverse set of functionalities.

In this section we compare our main goals with some of the tools already available. We focused mainly on tools that have more similarities with our approach. Therefore, we are looking mainly to tools that include more than one kind of task and targeting more than one type of user. Then, we looked up their popularity.

The main differences from the analysed platforms and our main goals are:

- some of the platforms are not NLP specific, like Mashape. They just work like a proxy that hides some of the web-services requirements (like user authentication and quota management). Nevertheless, there is no information about how the real service is implemented, and if its architecture is generic enough to be configured for other requirements;
- other platforms, like Text-Processing, although allow different types of services, all of them are based on one single tool (in this case, NLTK). Again, no information is given on the system implementation and how it can be adapted to other tools, and in specific, for functionalities not available in NLTK.
- and finally, mono-application services. Some are available together in a similar place, like CORE API by TextAlytics but there is no integration or homogeneity between the different offered services.

During the development of SplineAPI our main goal is to have an extensive system, to be used by anyone interested in offering Web Services, that can be easily configured and monitored.

3   Design Goals and Architecture Details

The main goal is to create a solution that minimizes the challenges developers face, when trying to take advantage from a large set of NLP tools already available.

In today's connected world, applications are no longer running only on the client machine. Also, they are no longer running only server-side. They are distributed, both on the client machine, server machine and others that might help in the process.

Therefore, our goal is to help the conversion of NLP tools into web services. Although the tool installation may be a challenge, the administrator of these services needs to deal with it, we intend to make the API construction easy, recurring to a set of Domain Specific Languages (DSL).

With the idea of creating a web API, it was necessary to think what is the best implementable architecture to develop this idea. The easiest and the cleanest method, to make available all the NLP tools, is to build a web service. Inside the web service *world*, there are various options of architectures, depending on how do we want to provide the service. The most popular are: Simple Object Access Protocol (SOAP) and Representational State Transfer (REST), each one with its own advantages and disadvantages depending on the objective in mind. When it comes to SplineAPI, the obvious choice was REST [2,8].

REST is more and more popular, and the best benefit it offers, is the optimization for stateless interactions that, in this case, is an essential feature, because the platform handles specific requests and responses based on text data, and that, does not require a connection status. To the users, REST is the simplest way to query a service because it is less verbose and easy to understand, as it bases its interaction with the clients in well known HTTP commands.

With the platform's architecture decided, it was then fundamental to investigate the best way of developing all the connections between the tools and the service, and the software technologies needed to make everything work.

3.1 Spline Architecture

Figure 1 shows our solution architecture. The server is composed of three main components: the Spline REST server, the NLP tools and their interface definitions, and a quota database.

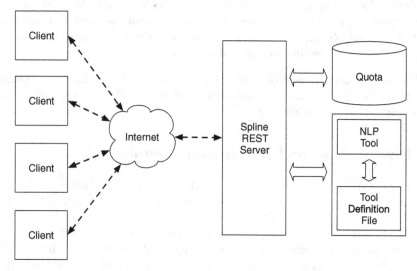

Fig. 1. Spline architecture.

NLP Tools and Definition Files. Different NLP tools communicate in different ways with the user. Some tools are command line applications that read information from a file, or from the standard input, and produce results in another file, or into the standard output. Some other are library-based, meaning that they expose an API that can be used in order to process information and obtain a desired output.

In order to be able to tackle with these different aspects of tools, each tool interface is described in an XML file.

This XML file is processed and a Perl module is created. This Perl module is responsible for the interaction with the Spline REST server, as is detailed in Sect. 3.2.

Listing 1.1. XML example for the Tokenization Service based on FreeLing Perl library.

```xml
<service>
  <meta batch=" false ">
    <tool>FreeLing</tool>
      <name>Tokenizer</name>
      <route>tokenizer</route>
      <parameters>
        <parameter required=" true " name=" text " type=" textarea ">
          <description>The text to be tokenized</description>
        </parameter>
      </parameters>
      <definition>Process of breaking a stream of text up into tokens
        .</definition>
      <cost>1</cost>
  </meta>
  <implementation>
      <packages>
        <package>FL3 'pt'</package>
      </packages>
      <main lang=" perl ">
        my $pt_tok = Lingua::FreeLing3::Tokenizer ->new(" pt ");
        my $tokens = $pt_tok->tokenize($text, to_text => 1);
        return $tokens;
      </main>
  </implementation
  <tests>
      <test>
        <param name=" text ">I will be tokenized .</param>
        <code>
          ok($result ->[0] eq 'I', " Test_the_first_word");
        </code>
        <code>
          ok((scalar @{$result}) == 5, " Test_the_result_length");
        </code>
      </test>
  </tests>
  <documentation>
    <header title=" module ">Spline::FreeLing::Tokenizer - a module that
        tokenizes your text .</header>
  </documentation>
</service>
```

The XML structure follows a proper XML Schema that allows the validation of the XML file. It also defines the domain of specific elements and attributes, which allow easy verification on the XML semantics.

Listing 1.1 presents an example of an XML definition file. It describes the interface for a tokenization service based on FreeLing [7] library.

The XML file is composed by three main parts:

- The meta-data for the service includes its name, the back-end tool and the service route (basically, the path used for the service URL). It also includes a brief explanation of the service goals, the service usage cost (if applicable) as well as which parameters should be used in order to request an operation. Each parameter is described in terms of its name, requiredness, data type (text, number, file and others) and default values. Note that, when a service receives file parameters, it request needs to be performed using the POST HTTP method, with multipart form data.

The file also includes documentation, adding a brief explanation of each parameter meaning.

– A description of how the parameters supplied by the users will be used to compute a result. At the moment this is done using Perl code or Bash commands. In the first case, there are two sections, one describing the Perl packages that need to be loaded, and another with the code that is executed. For Bash commands, only the executed code section should be used.

We are aware that for different tools our generator will have different needs, and therefore this section of the XML definition file might need further options in the future.

– It includes a set of tests that allow the service programmer or the server administrator to test if all services are working properly. These tests include an input for the service and a set of assertions over the obtained output. Again, at the moment these tests are being written directly in Perl, but we have been working into incorporate a JSON querying language like JsonPath[1] or JSONiq[2].

– Finally, the file adds the possibility to document the service as a module. It divides the information by headers (like chapters) and each one has its own proper description. It is simply a way to maintain the system's organization as well as explain everything in more detail.

The structure of the Perl module generated from these XML definition files is presented later, in Sect. 3.2.

Quota Database. Although our service is designed to be stateless, meaning that the service is connection-oriented, we want to record information on service usage, in order to track users, most used services, and if possible, distribute different services by different servers, so that highly used services are hosted in different hardware.

In one hand, each service defines how much a request to it costs. This cost can be a constant or defined accordingly with the amount of data to be processed. On the other hand, each client has an amount of quota to be used based on a cost limit. This quota can differ accordingly with the status of the client or, who knows, accordingly with a paid plan. Of course there is also the possibility to turn off quota management completely.

For this to be possible it was created a coin strategy. Each user has a daily limited amount of coins he can use freely. All the functionalities are different in their processing time but have a text-based parameter that can be small or big and, based on that, we stipulated a whole panoply of cost indicators that differ with the length of the text and the functionality itself. For that to happen, it was obviously fundamental to create a stateless authentication process to identify and manage all the users and their requests.

---

[1] A XPath like language for JSON, available from: http://goessner.net/articles/JsonPath/ (Last visited: 15-04-2015).

[2] A very complete and expresssive query language for JSON, available from: http://www.jsoniq.org/ (Last visited: 15-04-2015).

Spline REST Server. Considering that Perl is a programming language adequate to process textual data, with a great set of interfaces to other programming languages, it was the chosen language for the back-end server implementation.

The server is implemented in Perl, using the Dancer2 Web Framework [10]. The interaction with the NLP tools is done using Perl modules generated automatically from the already mentioned XML Definition Files. These modules are loaded automatically by the server, making all services available.

The server is responsible for querying the quota database and update it accordingly with the user requests. When called using the standard HTTP protocol, it presents common web pages documenting the services that are available (accordingly with the loaded modules) and their interfaces.

This strategy allows the easy creation of new services, just by creating an XML definition file, converting it into a Perl Module (and in some cases, some edition of the generated module) and restarting the web server. The new module will be loaded and its description and documentation will be made available in the website automatically.

3.2 Perl Module Generation

As already mentioned, the XML definition file is processed and "compiled" into a Perl module. The Perl module includes information about the service itself (namely, the `meta` section of the XML definition file) and a set of methods that are used both for configuring the service, and to perform the required operations to provide the service.

The module generation is template based. The meta-information is converted into an associative array (*hash*, in Perl terminology), and the Perl code is embedded in a subroutine.

The generated Perl module can be edited manually, to perform any special tweaks or improvements that might be necessary.

Listing 1.2 shows the relevant portions of the generated Perl module. Each module should implement a programming interface (called Roles, in Perl world), making available functions to access some of the needed data. Some of these functions have default behaviour, and as such, the code generator creates stub functions that can be then edited by the user. This means that the XML description tion can be used just for the module bootstrap.

The Perl module should also include a main function that will receive the request in a dictionary, and should return an answer as a Perl structure. This structure will be then converted into JSON and sent to the client.

In the Perl community a Perl module is, usually, shipped together with a set of tests. Therefore, the test information available in the XML definition file is used to generate such tests, like the one presented in Listing 1.3.

These tests can be used both for testing the Perl module locally, as well as to test the production service (in order to guarantee all the services are running correctly).

Listing 1.2. Module generated by the XML example.

```perl
package Spline :: FreeLing :: Tokenizer;

use FL3 'pt';

my %index_info = (
  hash_token => 'tokenizer',
  parameters => {
    api_token => {
      description => "The token to identify the user",
      required => 1,
      type => 'text',
    },
    text => {
      description => "The text to be tokenized",
      required => 1,
      type => 'textarea',
    },
  },
  description => "Process of breaking a stream of text up into tokens
        .",
  cost => 1,
);

sub get_token { return $index_info{hash_token} }

sub get_info  { return \%index_info }

sub cost_function{
  # return the total cost of the request
  return $index_info{cost};
}

sub param_function {
  # return 0 or 1 depending on the validation of the request
  return 1;
}

sub main_function {
  my ($input_params) = @_;
  my $result = _freeling_tokenizer($input_params);
  my $json = encode_json $result;
  return decode_utf8($json);
}

sub _freeling_tokenizer{
  my ($input_params) = @_;
  my $text = $input_params->{text};
  return unless $text;

  my $pt_tok = Lingua :: FreeLing3 :: Tokenizer->new("pt");
  my $tokens = $pt_tok->tokenize($text, to_text => 1);
  return $tokens;
}

1;
```

Listing 1.3. Tests generated by the XML example.

```perl
use strict;
use warnings;
use HTTP::Tiny;
use JSON;

use Test::More tests => 2;

my $host = $ENV{SPLINE_HOST} || 'localhost';
my $port = $ENV{SPLINE_PORT} || 8080;

my %params = ();
$params{api_token} = 'a_token';
$params{text} = 'I will be tokenized.';

my $got = HTTP::Tiny->new->post_form("http://".$host.":".$port."/
    tokenizer", \%params);
my $result = decode_json($got->{content});

ok($result->[0] eq 'I', "Test the first word");

ok((scalar @{$result}) == 5, "Test the result length");
```

3.3 Lengthy Requests

The previously presented architecture works great when the processes can be run on the fly. Unfortunately, a lot of Natural Language Processing tasks are slow, that would fire HTTP timeouts easily. Also, if concurrent users try to perform such tasks, the system will overload and be even slower (if not failing at once).

With a big range of lengthy services in the NLP area, a solution to deal with this kind of tasks was needed. It was, then, necessary to delineate a way to close the connection to the user and, after the desired process is complete, return the results.

The solution was the development a system daemon, that processes requests from a queue, and make the results available to the end-user. The algorithm is based on the following outline:

1. The REST server receives the request and detects if it is a lengthy one.
2. For lengthy requests, the server answers with a JSON that includes the URL where the answer, when ready, will reside. At the same time, it creates a task in the daemon queue, and a JSON file, accessible to the user, describing that the task is running.
3. At this point, the first HTTP connection is already closed.
4. When possible, the daemon unqueues the task and executes it, placing the resulting files in the folder created by the REST server for that effect.
5. The end-user will request, periodically, the JSON file, checking if its content changed. If so, check the URL where the results are, and fetch them. This

process might be a problem if the users check for the JSON file changes too often. Nevertheless, a simple GET request should be faster and lighter than having the processes running at the same time.

To differentiate these services from the common ones, the XML file describing the process accepts an extra attribute. This type of modules need to follow the outline above.

To make the daemon work there are four main folders, the first two private, the second two, public:

- The *logs* folder is used by the daemon to save information on each processed request. It allows the administrator to track the daemon activities, and debug them.
- The *queue* folder store files describing each process in the queue. They are organised by time, therefore allowing its use as a queue.
- The *json* folder save the JSON files with the information to be delivered to the user. When the process is running, these files show that the process is not complete. When it ends, its content changes to include the URL for the final resources.
- Finally, the *results* folder will store the output files, that are kept for a couple of days, to allow the user to collect them.

Periodically the *json* and *results* folders are cleaned for too old files.

4   Conclusions

In this document we present the architecture for a module-based server for REST services. The motivation for its development is the need to make NLP related operations available easily, without all the problems that comprise their usual configuration and installation.

Although the whole framework is ready and some services are already available (http://spline.di-um.org/) we are aware that different tools will dictate different problems to manage. In fact, we are already aware of some of the challenges we will face:

- At the moment, the lengthy services code part is not generic and it is mandatory that the admin manage a big part of the process. To improve this issue, it will be added an output section to the XML generation schema. This sections will contain all the URLs to the result files and it will be kept in the Perl Module to use when the final JSON file is created. With this feature, the user knows where to find the desired information before the request is completed (although it maintains a flag that informs the process is still running) and the admin does not have to deal with that.
- Turn the platform even more user friendly. There are some things that can be improved like error responses, interface organisation and styling and some specific features.

Other than these developing challenges we intend to implement in Spline, we will face other problems as soon as the server starts to be widely used, namely computational weight and server balancing.

Acknowledgements. This work has been partly supported by FCT - Fundação para a Ciência e Tecnologia within the Project Scope UID/CEC/00319/2013.

References

1. Cunningham, H., Maynard, D., Bontcheva, K.: Text Processing with GATE. Gateway Press, California (2011)
2. Fielding, R.T.: Representational State Transfer (REST). Ph.D. thesis, University of California, Irvine (2000). https://www.ics.uci.edu/fielding/pubs/dissertation/fielding_dissertation.pdf
3. Liu, B.: Sentiment Analysis: Mining Opinions, Sentiments, and Emotions. Cambridge University Press, New York (2015)
4. Loper, E., Bird, S.: Nltk: the natural language toolkit. In: Proceedings of the ACL-02 Workshop on Effective Tools and Methodologies for Teaching Natural Language Processing and Computational Linguistics, ETMTNLP 2002, vol. 1, pp. 63–70. Association for Computational Linguistics (2002)
5. Mani, I., Maybury, M.T.: Advances in Automatic Text Summarization, vol. 293. MIT Press, Cambridge (1999)
6. Martin, J., Jurafsky, D.: Speech and Language Processing: An Introduction to Natural Language Processing, Computational Linguistics, and Speech Recognition, 2nd edn. Prentice Hall, Upper Saddle River (2009)
7. Padró, L.: Analizadores multilingües en freeling. Linguamática **3**(2), 13–20 (2011)
8. Pautasso, C., Zimmermann, O., Leymann, F.: Restful web services vs. big'web services: making the right architectural decision. In: Proceedings of the 17th International Conference on World Wide Web, pp. 805–814. ACM (2008)
9. Rychtyckyj, N.: Machine translation for manufacturing: a case study at ford motorcompany. In: Proceedings of the 18th Conference on Innovative Applications of Artificial Intelligence, IAAI 2006, vol. 2, pp. 1728–1735. AAAI Press (2006). http://dl.acm.org/citation.cfm?id=1597122.1597130
10. Sukrieh, A.: Dancer2: Manual - A gentle introduction to Dancer2 (2013). http://search.cpan.org/sukria/Dancer2-0.10/lib/Dancer2/Manual.pod

Engaging Researchers in Data Management with LabTablet, an Electronic Laboratory Notebook

Ricardo Carvalho Amorim[(⊠)], João Aguiar Castro, João Rocha da Silva, and Cristina Ribeiro

INESC TEC–Faculdade de Engenharia da Universidade do Porto, Porto, Portugal
{ricardo.amorim3,joaoaguiarcastro,joaorosilva}@gmail.com, mcr@fe.up.pt

Abstract. Dealing with research data management can be a complex task, and recent guidelines prompt researchers to actively participate in this activity. Emergent research data platforms are proposing workflows to motivate researchers to take an active role in the management of their data. Other tools, such as electronic laboratory notebooks, can be embedded in the laboratory environment to ease the collection of valuable data and metadata as soon as it is available. This paper reports an extension of the previously developed LabTablet application to gather data and metadata for different research domains. Along with this extension, we present a case study from the social sciences, concerning the identification of the data description requirements for one of its domains. We argue that the LabTablet can be crucial to engage researchers in data organization and description. After starting the process, researchers can then manage their data in Dendro, a staging platform with stronger, collaborative management capabilities, which allows them to export their annotated datasets to selected research data repositories.

1 Introduction

With increasing amounts of research data being produced every year [3], institutions tend to implement guidelines and workflows to preserve them, in a similar way to what it is already the current practice with publications [6]. Nevertheless, this approach can pose some barriers to the dissemination and reuse of such datasets, as a consequence of the lack of metadata that is essential for other researchers to understand the production context of a specific dataset [12]. Likewise, gathering domain-level metadata at the deposit stage can be a very demanding—and time consuming—task for curators, often responsible for more than one research domain. At the same time, researchers play a key role in their data description [7], as they have the best knowledge of their production environment, and can add metadata to their data from the early stages. Existing platforms for research data management, such as Figshare[1] or Zenodo[2], already

---

[1] http://figshare.com/.
[2] http://zenodo.org/.

© Springer International Publishing Switzerland 2015
J.-L. Sierra-Rodríguez et al. (Eds.): SLATE 2015, CCIS 563, pp. 216–223, 2015.
DOI: 10.1007/978-3-319-27653-3_21

support simple descriptive metadata, but the barrier between them and the researchers' working environment is still high. It is therefore recognized that important data and metadata are still temporarily stored in frail locations such as personal computers and laboratory notebooks [10]. Ultimately, even with guidelines for data management in place, some of these resources never reach the deposit stage as they are susceptible to neglect.

In this paper, we present LabTablet as an application to help researchers gather data and metadata during experimental runs or field trips, and directly export them to a staging repository—in our case Dendro [5]—responsible for creating a collaborative, description-oriented approach to research data management. With this approach, we can provide a better handling of research data and provide conditions for capturing metadata as soon as it becomes available. At the end of a research project, Dendro is capable of creating and exporting the dataset package to existing platforms for data preservation, that can also take advantage of the included metadata to improve the visibility of the dataset.

2   Research Data Management

Amid the research activities, researchers produce both raw and processed data that support their conclusions towards the project goals. These resources are sometimes neglected after the publication of the results, weakening the link between project results and the data that supported them.

Managing research data has evolved to include tasks besides storage and preservation, ensuring a proper handling of research outputs to facilitate their retrieval and long-term preservation. Furthermore, similarly to what happens with research publications, the deposit of research assets in repositories has to be accompanied by a comprehensive description—also known as metadata records—to facilitate their retrieval and interpretation. Ideally, when a dataset is provided with sufficient metadata, others will be able to reuse it [12]. An equally measurable result is the credit that researchers get from publications citing their data, with side effects related, among others, with the possible reduction of costs inherent to the research activity.

2.1   Data Description

Datasets and publications have different requirements concerning their description [11]. Considering the diverse scenarios in which datasets are produced, we can identify sets of possible metadata descriptors that can be directly related to each specific research domain, and at the same time extend the basic, high level ones, used to describe publications. For each research domain, the description possibilities vary, and thus, the data repositories are evolving to comply with this required flexibility [2].

Well-known metadata schemas, such as Dublin Core, have been considered fit to a broad scope of applications and allowed the emergence of protocols for exchanging metadata and enhancing publications visibility [9]. The OAI-PMH[3] is the best known, and is widely used to index different repositories, allowing their resources to be presented in publications search engines. Basic descriptors, such as title, description and author, can be added by a designated curator and provide the link between data and publications, but when considering the broad possibilities for description in each of the domains, this task has to include researchers. Actively involving researchers in the description of their data faces some limitations, as the platforms created for this purpose must also take into account usability requirements and offer features that meet their goals as researchers, such as receiving credit for their data and sharing them with their peers.

2.2   Researchers' Engagement in Data Management

In the course of research activities, researchers often resort to personal computers to store collected data and to their laboratory notebooks to record any observations or context. With the increasing amounts of research data, these approaches pose some risks in terms of data preservation, which can later constrain data's availability.

In the past few years, several platforms emerged to integrate the research environment, with some of them being actively used by several communities [2]. These platforms aim to implement established protocols for data preservation and dissemination, while featuring easy to use interfaces along with collaborative environments. The assessment of several existing platforms showed that issues such as data ownership, dataset description and dissemination are already a concern, although these platforms are still considered as a final location for dataset deposit [2]. Staging platforms such as Dendro, on the other hand, aim at creating management tools closer to the researchers' daily routines and offer a place where they can collaboratively store and describe data. It is important to stress out that, for these platforms, all the managed data are private and unaccessible from the external community, as it can involve sensitive data that have to be adjusted prior to its disclosure. Only then they can be cited and reused. At the end of the research activity researchers can export the resulting resources to the final repositories, often aimed at long-term preservation.

3   Electronic Laboratory Notebooks

We have previously highlighted the importance of data management repositories, both as staging environments and as research data preservation solutions. As several researchers resort to field trips or experimental runs to gather data—often a typical approach to data production—there is still a gap between data production and their deposit in the mentioned platforms. Electronic laboratory notebooks can fill this gap, allowing researchers to record and directly deposit

---

[3] https://www.openarchives.org/pmh/.

data, while mitigating the risk of loosing such records during the process [8]. Nevertheless, the existing solutions tend to focus on a particular domain or offer limited functionality, not taking advantage of some of the available sources of metadata, and excluding prospective users from other domains.

3.1   LabTablet

Taking advantage of the growing popularity of handheld devices, LabTablet was developed as an electronic laboratory notebook to help researchers describe their data as soon as the project starts. Besides having an easy to use interface, the underlying representation for each metadata record follows established standards, ensuring a streamlined curation process before the final deposit in a repository. The first version of this project was focused on gathering metadata in the field, relying on previously built application profiles, and therefore using a set of descriptors for that specific domain. In any of the versions, LabTablet is capable of uploading each dataset to Dendro, or any other staging platform, from which it can later be included in preservation solutions. This approach allows curators to have standards-compliant metadata records upon deposit but, more importantly, domain-level metadata that would otherwise be lost is properly maintained.

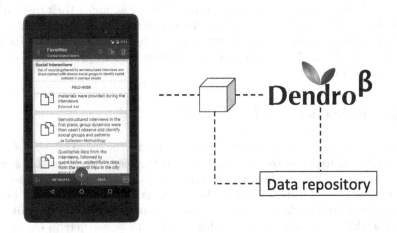

Fig. 1. View of a project with the gathered metadata.

Figure 1 the application's interface regarding an opened project, with the correspondent gathered descriptions that can later be exported to a designated platform. After preliminary evaluation with researchers from the biodiversity domain [1], a new approach was developed, extending the metadata capabilities of this application and including mechanisms to also gather opportunity data—observations collected by chance while performing some other activity. Opportunity data can be directly linked to the researchers' field trips and be enriched by the use of the tablet's built-in sensors to gather metadata from the available sources such as camera, GPS or accelerometer. In addition to those, LabTablet also allows voice recordings, sketches, and tracking a field trip, and is able to export the results to a compliant format[4]. Furthermore, researchers can also import other types of data (namely spreadsheets) from their computers, merging them into the workflow. To take advantage of the device's capabilities, and considering a wider set of research domains, additional input modes were also implemented, namely forms, used in surveys. Forms can be custom-designed and filled directly in the application. The workflow for such process relies on the researcher to create a model, and to instantiate it whenever a subject is interviewed. The same applies for other activities that require some kind of form or survey such as routine evaluations and observations. The gathered data is then exported to files that are compatible with common statistical analysis tools such as Excel or SPSS[5]. At this stage, the development of new LabTablet features is mainly dependent on the integration of existing workflows, as well as the inclusion of standards that are already in use among the research workgroup or possible direct connection between the application and the researcher's tools (LabView[6] or SPSS, for instance).

At the end of each field trip (or when the researcher finds it convenient to do so), the application can sync the collected resources with a repository where researchers are able to share them with their team or the community. This ensures that data are stored in the appropriate location, under their institutions supervision. Additionally, as with metadata records, the created package can follow any guidelines, namely the structure of a Submission Information Package (SIP), from the Open Archival Information System model[7], provided the correct integration is done.

4   Social Sciences: A Case Study

As a part of an ongoing partnership, a researcher from the social sciences domain was interviewed to assess the different data management needs for this specific domain[8]. Initially, a set of questions was proposed to address metadata needs or possible constraints on data sharing. As the interview went on, several important aspects related to the workgroup's current practices allowed us to tailor the existing workflow to their needs. Previous work with researchers in engineering domains [4] showed that usually researchers deal with systematic data production which has features that are common to several domains: experimental data, for instance, tends to deal directly with the experimental setup and the physical

---

[4] A KML-based representation (https://developers.google.com/kml/), containing a set of connected coordinates, for instance.

[5] http://www-01.ibm.com/software/analytics/spss/.

[6] http://www.ni.com/labview/pt/.

[7] http://www.iso.org/iso/catalogue_detail.htm?csnumber=57284.

[8] The survey for this evaluation was based on the Data Curation Toolkit, available at http://datacurationprofiles.org/.

properties of samples or compounds. In the social science domains, on the other hand, workflows are centered on temporal or spatial coverages, having their main focus on social traits that can differ greatly. As a result we have high heterogeneity of dataset structures and description needs across different research groups, that are highly dependent on the researcher's view of the event.

4.1 The Social Sciences Domain

Our interview revealed the researcher's awareness of the recent evolution of data management guidelines on this area. However, these had never been put into practice. Studies in this group are mainly focused on evaluating phenomena in different social groups, directly interacting with them either through field observations, structured or unstructured interviews, or content analysis. During these activities, the produced data is mainly of qualitative nature, with a small portion of quantitative data as well. Qualitative data is, for this group, mostly related with observations or notes which contents are fully dependent on the producer, whereas quantitative data results from surveys and questionnaires.

Concerning the publication of research data, the researcher highlighted some limitations, as some projects are not expected to disclose data and some datasets are of a sensitive nature and need to follow ethical recommendations, or need to be anonymized before their disclosure, if applicable. Still, for some projects, pursuing data disclosure would benefit both parts, as they would be able to cite datasets in publications and their peers could access and reuse such data in subsequent analysis.

LabTablet proved to be capable of handling all these needs in terms of data production, as well as helping researchers identify some equally important descriptors that could be added to provide extra context. During the course of this interview, a set of basic Dublin Core elements revealed to be satisfactory for the description needs in this domain[9]. Nevertheless, for a deeper data description, other schemas should also be included to achieve an extensive metadata record.

4.2 Preparing for Data Description

After identifying the basic description needs and suggesting an initial profile for this purpose, we proceeded to identify other domain-level descriptors. In this field, the Data Documentation Initiative (DDI)[10] proved to have a suitable set of descriptors for social sciences domains, namely[11]:

- **Data Collection Methodology**—to specify which methodology was used to collect the samples or questionnaires. This revealed to be a recurrent scenario as researchers often worked with a small set of methodologies;

---

[9] These consist of the base Dublin Core elements profile, namely abstract, contributor, creator, subject, title, description, publisher, date, type, and others, as specified in http://dublincore.org/documents/usageguide/elements.shtml.

[10] http://www.ddialliance.org/.

[11] Not all the descriptors are depicted here.

- `Data Source`—to identify the source of the collected data, including the associated project. As some of the projects could include partnerships with other data providers, this descriptor was chosen to support such specification;
- `Sample Size`—to state the dimension of the sample or the number of interviewees during a field session;
- `External Aid`—a reference to any support given during the experiment, such as text cards or multimedia support;
- `Kind of data`—a specification of the dataset's content type. This allows researchers to specify whether the packaged data is of a qualitative, quantitative or mixed type;
- `Universe`—a description of the referenced population, if applicable. This can include informations related to age, gender or income classifications.

The selected descriptors allow a better understanding of the dataset in question. A clear description of the population will, for instance, enable other researchers to search for datasets that were obtained from specific social communities, and the same happens for the other descriptors such as the `Sample Size`. According to the researchers, identifying the methodology was considered to be a key item in the description process. This identification was often extensively done and it was a common item to be mentioned in each project. According to the schema specification, this item is expected to mainly consist of a brief description of the involved methodology, but in this case—and considering related work in this area—this field can sometimes be very extensive, which led the researcher to suggest that other descriptors should also be present to promote a structured representation of this information.

After this selection of descriptors, we proceeded to create the ontology for this domain. Along with the descriptors from the Data Documentation Initiative, we included high level descriptors from the Dublin Core profile as well. This ontology can be loaded at any time into the LabTablet application and be used to describe data in this area. The same is true for Dendro, our staging repository.

5   Conclusions

By analyzing different research domains, we can identify many differences concerning data management practices. While some groups have data management procedures already in place, most are still far from addressing the issue, mostly due to the nature of their data rather their motivation.

The researcher from our case study recognized the added value in automatically exporting the daily produced data to a centralized location, where it could be properly handled and edited. Additionally, some specialists in the field advise against using any kind of note taking tools during the interviews, not to influence the interviewee; however, the researcher considered very important to be able to record or transcribe the interviews in the background.

We are testing the collection of metadata throughout the entire research workflow with several research teams. It is clear by now that devices and tools to make the process easier on the researchers can make the difference between a process regarded as an extra burden on researchers and one where they perceive the benefits and get involved.

Acknowledgements. Project SIBILA-Towards Smart Interacting Blocks that Improve Learned Advice, reference NORTE-07-0124-FEDER000059, funded by the North Portugal Regional Operational Programme (ON.2-O Novo Norte), under the National Strategic Reference Framework (NSRF), through the European Regional Development Fund (ERDF), and by national funds, through the Portuguese funding agency, Fundação para a Ciência e a Tecnologia (FCT). João Rocha da Silva is also supported by research grant SFRH/BD/77092/2011, provided by the Portuguese funding agency, Fundação para a Ciência e a Tecnologia (FCT).

References

1. Amorim, R.C., Castro, J.A., da Silva, J.R., Ribeiro, C.: LabTablet: semantic metadata collection on a multi-domain laboratory notebook. In: Closs, S., Studer, R., Garoufallou, E., Sicilia, M.-A. (eds.) MTSR 2014. CCIS, vol. 478, pp. 193–205. Springer, Heidelberg (2014)
2. Amorim, R.C., Castro, J.A., Silva, J.R., Ribeiro, C.: A comparative study of platforms for research data management: interoperability, metadata capabilities and integration potential. In: Rocha, A., Correia, A.M., Costanzo, S., Reis, L.P. (eds.) New Contributions in Information Systems and Technologies. AISC, vol. 353, pp. 101–111. Springer, Heidelberg (2015)
3. Borgman, C.L.: Advances in information science: the conundrum of sharing research data. J. Am. Soc. Inf. Sci. Technol. **63**(6), 1059–1078 (2011)
4. Castro, J.A., da Silva, J.R., Ribeiro, C.: Creating lightweight ontologies for dataset description. Practical applications in a cross-domain research data management workflow. In: IEEE/ACM Joint Conference on Digital Libraries (JCDL), pp. 313–316, London (2014)
5. da Silva, J.R., Castro, J.A., Ribeiro, C., Lopes, J.C.: The Dendro research data management platform: applying ontologies to long-term preservation in a collaborative environment. In: Proceedings of the iPres 2014 Conference (2014)
6. Lynch, C.A.: Institutional repositories: essential infrastructure for scholarship in the digital age. Association for Research Lybraries. Bimonthly Report no.226 (2003)
7. Lyon, L.: Dealing with data: roles, rights, responsibilities and relationships. Consultancy Report, UKOLN, pp. 1–65, June 2007
8. Jason, T.: Nickla and Matthew B Boehm. Proper laboratory notebook practices: protecting your intellectual property. J. Neuroimmune Pharmacol. **6**(1), 4–9 (2011)
9. Rice, R.: Applying DC to institutional data repositories. In: Proceedings of the International Conference on Dublin Core and Metadata Applications, pp. 212-212 (2008)
10. Tenopir, C., Allard, S., Douglass, K., Aydinoglu, A.U., Wu, L., Read, E., Manoff, M., Frame, M.: Data sharing by scientists: practices and perceptions. PLoS ONE **6**(6), e21101 (2011)
11. Treloar, A., Wilkinson, R.: Rethinking metadata creation and management in a data-driven research world. In: IEEE Fourth International Conference on eScience, pp. 782–789 (2008)
12. Willis, C., Greenberg, J., White, H.: Analysis and synthesis of metadata goals. J. Am. Soc. Inf. Sci. Technol. **63**(8), 1505–1520 (2012)

OFR: An Efficient Representation
of RDF Datasets

Jakub Swacha[1][✉] and Szymon Grabowski[2]

[1] Institute of Information Technology in Management, University of Szczecin,
Mickiewicza 64, 71-101 Szczecin, Poland
jakubs@uoo.univ.szczecin.pl
[2] Institute of Applied Computer Science, Lodz University of Technology,
Al. Politechniki 11, 90-924 Łódź, Poland
sgrabow@kis.p.lodz.pl

Abstract. The constant growth of structured data, often in the form
of RDF, demands for efficient compression methods, to facilitate their
storage and transmission. We propose an RDF compression algorithm
that produces a succinct representation of RDF datasets. It consists of
two stages. The first splits the input triples into multiple streams, and
applies tailored compaction techniques for each stream. In the second,
a general-purpose compression is applied. We experimentally show on a
number of datasets that the proposed algorithm achieves compression
ratios significantly better than the RDF compressors known from the
literature.

1 Introduction

The Resource Description Framework (RDF) is a family of W3C specifications
used for modeling information on the Web. RDF is a vital part of the Semantic Web vision in which ultimately all information on the Internet is machine-
processable, shifting the (still dominating) document-centric perspective to the
data-centric one. Although still in infancy, large RDF datasets appear and grow
at a fast pace, currently containing in total over 25 billion triples (http://www4.
wiwiss.fu-berlin.de/lodcloud/).

RDF statements have a subject, a predicate (also known as a property) and
an object term. The domain of subjects and objects are typically partially shared.
Note that an RDF dataset may be perceived as a directed labeled graph, with
possibly more than a single label (or labeled edge) between a subject and an
object.

Some RDF domains have been adopted, for example, in life sciences
(e.g., Uniprot), geography (e.g., Geonames), sensor data from weather sta-
tions (LinkedSensorData) and open government (e.g., U.S. data.gov, U.S. Cen-
sus data). The result of a notable effort is also DBpedia, a dataset containing
extracted data from Wikipedia, with about 2.6 M concepts described by 247 M
triples.

J.-L. Sierra-Rodríguez et al. (Eds.): SLATE 2015, CCIS 563, pp. 224–235, 2015.
DOI: 10.1007/978-3-319-27653-3_22

This growth raises challenges and requires a succinct representation of the RDF repositories, with two goals in mind: possibly small compressed size for distribution and exchange, and possibly small representation supporting queries. Originally (in 1999) XML syntax for RDF data model was recommended but other serialization choices are also in use and even gain popularity. One of these alternatives is N-Triples (http://www.w3.org/2001/sw/RDFCore/ntriples/), a line-based plain-text format.

The outline of this article is as follows. In the next section we give a brief outline of the RDF compression area. Section 3 presents our algorithm, and its experimental results are shown in Sect. 4. The last section concludes.

2 Related Work

RDF compression is a relatively young research topic. Early approaches map RDF to a relational database. This can be achieved basically using one of three possibilities. The simplest and most natural is just to store all triples in a single 3-attribute table [5], an approach known under the name of a triple-store and used in the RDF storage systems Jena, Oracle, Sesame and 3store. Since URIs and literals tend to reoccur, and are relatively long, many such solutions do not store entire strings in the table, substituting them with their shorter versions or mapping them to numerical IDs. Still, in [1] it was shown that triple-stores hardly scale if the number of triples exceeds 50 millions.

An alternative, known for faster data access, are *property tables*. There are two variants of this approach. In the *clustered property tables* variant, several tables are built and the attributes (columns) in each are properties common to the triples stored in those tables. The left-over triples are stored in a triple table. In a second variant, a *property-class table*, the type property of subjects is used to cluster similar sets of subjects. Property tables can reduce subject-subject self-joins of the triples tables, yet if a query requires combining data from several tables, they become problematic [1,19]. Abadi et al. [1] point out other issues with property tables: many NULL values in the tables (since real RDFs are not very structured or complete) and the (surprising) abundance of multi-valued attributes, even some that seemingly should be single-valued (title of a book). The former deteriorates the performance, the latter requires storing multi-valued properties as e.g. lists (together with other attributes in the same table), which complicates query handling.

Yet another RDBMS-based approach, called *vertical partitioning*, is to have one table per one property [1]. In other words, its idea is to group triples by predicate and thus to obtain many 2-attribute tables, one for each predicate value (in many cases the number of used predicates is indeed small, e.g. less than 200).

Other works should also be mentioned: Hexastore [18] and RDF-3X [15] systems create indexes for all six element ordering combinations. RDF-3X additionally applies gap compression in leaves of the underlying B^+-tree, to make the indexes more compact. BitMap [3] also applies gap compression, for 2D matrices: SO and OS for each predicate, PO for each subject and PS for each subject.

All the mentioned schemes are not particularly succinct. Much better results have been achieved in two works from the same team. In [8] a compression scheme without random access to data (although with some random access friendly helper structures) was presented, grouping triples with the common subject to adjacency lists, and storing ordered object IDs for each property value bounded with that particular subject. The obtained sequences are then encoded statistically (with Huffman encoding as a major component) and the dictionary of strings (for S, P, and O values) is encoded with PPMd[1].

Finally, in [2] a scheme offering both good compression and fast query handling was proposed. This is based on the k^2-tree [4] data structure, being a pointerless variant of the well-known quad-tree. In the RDF case, for each property value a single k^2-tree is kept, representing a binary matrix of size $|SO| + |S - SO| \times |SO| + |O - SO|$, where SO is the set of common subjects and objects. Note that this matrix is rectangular (rather than square) but because of the partitioning nature of the k^2-tree, along both dimensions 0-cells will be padded, to the nearest power of k.

As the graph structure and the dictionary of strings are completely different components, no wonder that works focusing on a single aspect of RDF compression also exist. Martínez-Prieto et al. [14] consider efficient RDF dictionary compression. They consider several variants, more compression or more access time oriented, and utilize algorithmic techniques like hashing, front coding and FM-indexing, to name a few. Other researchers in this area focus on scalable parallel solutions [6,17], e.g. using the MapReduce paradigm.

Considering the graph structures, recent years also brought a few novel concepts. Joshi et al. [12] managed to losslessly prune over 50 % of the original triples for several popular datasets, using the idea of inferring triples from a set of logical rules derived from the given dataset. To give a toy example, if the dataset contains the triples $\langle A, fatherOf, B \rangle$, $\langle B, fatherOf, C \rangle$ and $\langle A, grandfatherOf, C \rangle$, the last of them may be inferred from the previous two, given an ontology explaining the connection between the relations "fatherOf" and "grandfatherOf".

A similar approach was taken by Pan et al. [16], who replace a frequently occurring graph pattern with a generalized triple. To give an example, if the pattern $\langle ?x, a, foaf : Person \rangle$, $\langle ?x, a, dbp : Person \rangle$ appears often (as it actually happens in the DBpedia dataset), a type T may be introduced, along with a rule to expand T to foaf: Person and dbp: Person. In this way, the single triple $\langle ?x, a, T \rangle$ would represent the pattern above.

Jiang et al. [11] proposed two algorithms. In one, they label each object and subject in the RDF graph with "type" and reduce the number of nodes by combining those with the same type and related neighborhoods. In the other variant, they contract the graph by removing nodes having only one neighbor after passing the information about the node to remove to its neighbor.

---

[1] http://www.compression.ru/ds/ppmdj1.rar (PPMd, var. J rev. 1, May 10, 2006, by D. Shkarin).

Hernández-Illera et al. [10] exploit the so-called predicate families, i.e., possible pairs of predicates and subjects. As in real datasets their number is much less than the combinatorial product of all existing values, there is a clear redundancy which may be removed. Similarly, there exists a redundancy in pairs of predicates and objects. The proposed algorithm, HDT++, produces archives for large datasets by 10–13 % smaller than the k^2-triples compressor [2].

A good survey to graph compression, including RDF compression, was recently published by Maneth and Peternek [13].

3 The Proposed Algorithm

The proposed algorithm is aimed at semantically lossless compression of RDF datasets. Our only goal was to improve compression efficiency, which would help to distribute large RDF datasets. Therefore, we do not aim at providing the capability of search without decompression, however most of the solutions presented here could also be applied to a search-capable compression scheme, and investigating this opportunity will be our future work.

3.1 The General Approach

The following list sums up the key characterics of our approach.

1. Separation of semantic and statistical encoding. The RDF-specialized transformations are performed as the first stage, whose output is compressed in the second stage with a general-purpose compression algorithm.
2. Separation of graph and dictionary compression. The graph contains only numeric identificators, which can be translated to the actual subjects, predicates, and objects using dictionaries. It was chosen to reduce the design complexity of the algorithm.
3. Splitting the content. The RDF dataset components that are of a distinct semantic type (e.g., subjects vs. objects) or consist of similar values (e.g., unmatched string fragments vs. match lengths) are put into separate streams. This helps to obtain more skewed distributions and allows patterns to form, which can be exploited by the second-stage algorithm, which compresses the separate streams, and outputs a single distributable file.
4. Reordering the content. The aim is to form a pattern that could be effectively encoded. The primary examples are: sorting the triples (so that at least the first triple component forms a non-decreasing sequence, thus becomes extremely prone to delta encoding), putting the triples in (object, subject, predicate) order (so that the highest entropy component becomes the one to be most effectively compressed after sorting with delta/run-length encoding), moving the numbers to the end of string (so that longer matches can be found as the numbers are often the most randomized part of strings).
5. Effective number encoding. The numbers represented in input formats as text are encoded in binary form, using all bits available in every byte.

6. Exploiting local redundancy. As similar values tend to appear next to each other in respective streams, we apply delta and run-length encoding to remove this redundancy or make it more prone to second-stage compression.

We called our approach Objects-First Representation (OFR), because of the order of triple elements.

3.2   Phases of the Algorithm

As noted above, the proposed algorithm consists of two stages: (1) semantic encoding of the RDF dataset content, and (2) statistical encoding and merging of the stage 1 output. As stage 2 simply involves the use of a general-purpose compression algorithm, below we shall only describe the phases of stage 1.

Parsing Input RDF Dataset. The first phase decodes the RDF dataset from its input representation. For every input line, first it checks its correctness. It is not a strict check, however errors that would cause failure of the subsequent compression phases are detected. Minor errors (e.g., unescaped UNICODE characters) are corrected, lines containing major errors (e.g., missing URI closing bracket) are put into separate *errorlines* output stream. Also, lines that do not contain triples (notably prefix definitions and comments) are put into separate streams. The lines considered correct are parsed into subject, predicate and object parts. The numbers are found and moved to the end of string, with their original places of occurence marked with zeros. The language and datatype information (sometimes appended to object literals) is extracted and replaced with a numerical identifier, with the original moved to a respective dictionary. The triple elements are then classified depending on their content type (*defined name*, *URI*, or *literal*), and queried in a respective dictionary (there are nine combinations of element and content types, but only six are valid in RDF, hence there are six main dictionaries; the two additional ones are for languages and datatypes). If an element is not found in a given dictionary, it is added, and has a new unique ID assigned. The three elements' IDs form a triple of numbers, which is appended to a list.

Sorting. All the main dictionaries are sorted lexicographically. New IDs are assigned to every dictionary item that (*i*) reflect the sorted order and (*ii*) ensure that the IDs for every element type are unique (e.g., no object literal has the same ID as any object URI). The triple list is updated by replacing all the old IDs with the new ones. Then, the triple list is sorted in the increasing order determined by object ID, then subject ID, then predicate ID.

Encoding Triples. The object ID's are encoded as run-lengths of series of zeros (denoting a number of triples referring to the same object) or ones (denoting a number of different objects). This is all that is needed for decoding, as the triples are sorted by object ID, therefore the consecutive objects in the triple list can only differ by having an ID increased by one. The subject IDs are encoded differently, depending on whether it is a first encountered subject for a given object (*leading subject*), or not (*consecutive subject*), to better exploit the non-decreasing order of the *consecutive subjects* introduced by sorting. Each but the very first *leading subject* ID is compared to the preceding one. If the difference is small, it is encoded using a single byte to the *subjects.hi* stream, otherwise, a range identifier is encoded using a single byte to the *subjects.hi* stream, whereas the remaining bits (pointing to the position within the range) are encoded to the *subjects.lo* stream. The *consecutive subject* IDs are delta- and run-length-encoded into three streams (*subjects.delta* for prefixes, *subjects.rle* for run-length subranges, *subjects.lo* for remainders). The predicate IDs are simply put to the *predicates* stream using a minimum possible number of bytes (taking into consideration the maximum predicate ID). Delta and run-length encoding was also tried with the *predicates* stream, and although it considerably reduced its size, the final effect, after applying second-stage compression, was found to be negative on test datasets.

Encoding Dictionaries. Each of the six main dictionaries is encoded in the following steps:

1. A dictionary element is matched to the preceding one. The match length is delta-encoded and put to the *matchlens* stream.
2. The unmatched part (including number markers but not including the numbers themselves) is put to the *dict* stream.
3. The number buckets (see steps 4 and 5) that were kept for offsets within the unmatched part are closed, with their content flushed into the *deltas.hi* and *deltas.lo* streams. Note that this grouping of numbers and delaying their output until a sequence is finished helps patterns to form and increases local redundancy, which is to be exploited by the second-stage compression.
4. The numbers that were encountered in the unmatched part (*leading numbers*) are separated into prefix (range identifier) and remainder (range position) parts, and put, respectively into the *nums.hi* and *nums.lo* streams. There are special prefix ranges reserved for numbers around 2000 and digits preceded by a single zero (e.g. 01), aimed at encoding months, hours, and popular years in a single byte. Moreover, encountering a number causes a new bucket to be initialized. The bucket will contain numbers encountered at the same offset in subsequent matches (see step 5).
5. The numbers that were encountered in the matched part (*consecutive numbers*) are delta- and run-length-encoded into a respective bucket of numbers, again with prefix and remainder separated. There are special prefix ranges reserved for 1000 and further powers of 10, so that such incrementals can be encoded in a single byte.

Because of their rather negligible size, the additional dictionaries (*languages* and *datatypes*) are passed to the second-stage compressor in their original format.

Decompression is a much simpler procedure. First, the additional dictionaries are read. Then, one after another, the main dictionaries are decompressed in three steps: (1) strings (without numbers) are reproduced using the content of the *matchlens* and *dict* streams, as well as the additional dictionaries; (2) the strings are analyzed to obtain number bucket sizes; (3) the numbers are decoded and inserted into places held by markers. Next, the triple element lists are decoded: (1) the object ID list; (2) the subject ID list; (3) the predicate ID list. Then, the lines that did not contain triples are copied to the beginning of the output (decompressed) file. Finally, the original triples are reconstructed by replacing the IDs with respective dictionary items, and appended at the end of the output file. Note that the order of triples is not preserved, which is not a problem, as the order of lines in the N-Triples format is irrelevant. This is why we called our scheme semantically lossless.

Figure 1 depicts, in a simplified way, splitting of RDF data into different streams, whereas Fig. 2 shows handling of exemplary RDF data.

3.3   Implementation Details

A proof-of-concept implementation of the proposed algorithm has been developed in C++ with the following design decisions:

1. The *openhash* of Zilong Tan's *ulib* library[2] was used to implement the main and additional dictionaries. This solution was found experimentally to handle large dictionaries much faster than the *hash_map* of the Standard Template Library.
2. The Standard Template Library's *sort* has been used for sorting both the triples and the dictionaries.
3. The object encoder uses the following range widths: 80 for prefixes of run-lengths of zeros, 32 for prefixes of run-lengths of ones, 16 for prefixes of run-lengths of (zero, one) pairs, and 128 for the remainders.
4. The *consecutive subject* encoder uses the following range widths: 32 for prefixes of run-lengths of zeros, 16 for prefixes of run-lengths of ones, 15 for small numbers (no remainder), 128 for numbers with a remainder one byte long, 64 for numbers with a remainder two bytes long.
5. The *leading subject* encoder uses ranges dependent on the number of subjects. The width of the range for prefixes is the minimum number that allows to encode the remainder with a static length code using a minimum number of bytes (multiples of 8 bits). The remaining codespace is used for delta coding (half for small numbers, the other half for prefixes of larger numbers, with a remainder one byte long).
6. Match length delta encoder writes them as bytes using (repetitions of) one special value (255) to encode values larger than 254.
7. The *leading numbers* encoder uses the following range widths: 112 for small numbers (no remainder), 40 for numbers with a remainder one byte long, 31 for numbers with a remainder two bytes long, 6 for numbers with a remainder

---

[2] Z. Tan, ulib. An efficient library for developing high-performance and scalable systems in C and C++, 2012, http://code.google.com/p/ulib/.

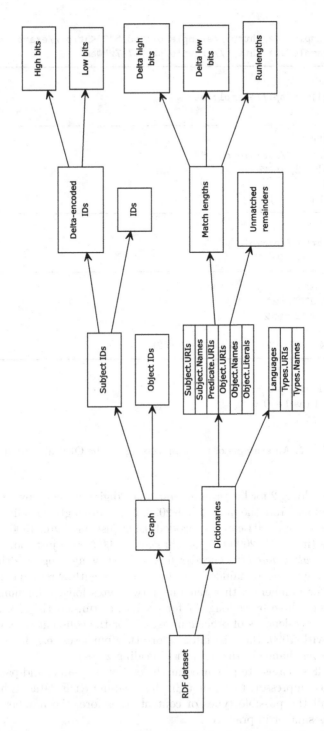

Fig. 1. General scheme of splitting RDF data by the OFR algorithm

Input triples:
`<http://example/s1> <http://example/p> "o3_5"^^<http://example/dt> .`
`<http://example/s2> <http://example/p> "o7_7z"@en .`

Dictionaries:
Datatypes (1): `<http://example/dt>`
Languages (1): `en`

Subject URIs:
Items (2): `"http://example/s0",""`
Match lengths (2): `0, 17`
Numbers (1): `1`
Deltas (1): `1` (as $2 - 1 = 1$)

Predicates:
Items (1): `"http://example/p"`
Match lengths (1): `0`

Object literals:
Items (2): `"o0^0", "z@0"`
Match lengths (2): `0, 2`
Numbers (1): `3, 5`
Deltas (1): `4` (as $7 - 3 = 4$), `2` (as $7 - 5 = 2$)

Triples:
Objects: `0, 1` (as $1 - 0 = 1$)
Subjects: `0, 1` (as $1 - 0 = 1$)
Predicates: `0, 0`

Fig. 2. An example of triple processing by the OFR algorithm

three bytes long, 2 for larger numbers, 10 for digits preceded by a single zero, 52 for numbers from the range 1969–2020, 2 for run-length encoding of leading zeros. Thus, e.g., 2004 can be encoded using just one byte, 10005 using just two bytes (in total), 2000007 just three, and 100000009 just four.

8. The *consecutive numbers* encoder uses the following range widths: 128 for small numbers (no remainder), 47 for numbers with a remainder one byte long, 31 for numbers with a remainder two bytes long, 6 for numbers with a remainder three bytes long, 17 for prefixes of run-lengths of zeros, 17 for prefixes of run-lengths of other numbers, 5 for the consecutive powers of 10, starting with 1000, the remaining 5 for other purposes (e.g. larger numbers and marking change in the number of leading zeros).

9. At most 48 streams are produced at the end of stage one, and passed to the stage-two compressor. Note, though, that real-life RDF datasets hardly ever contain all the possible types of content, therefore, the number of output streams is smaller in practice.

4   Experimental Results

We tested our C++ implementation of the proposed compression scheme (OFR) on a number of datasets in N-Triples serialization format. Table 1 presents compression ratios (in per cent) of several compressors. Results for both HDT and our OFR implementation were obtained experimentally with two backend compression algorithms: Deflate (as implemented in zip, using mode -9: max. compression) and LZMA (the default compression algorithm used in the 7z archiver, http://www.7-zip.org), the results of other prominent RDF compressors were copied from the available literature. As Table 1 shows, the proposed OFR algorithm achieves clearly best compression ratios across all test datasets.

Table 1. Compression ratio comparison. The results are given in per cent of the original dataset size. The results of HDT on the Mix dataset, marked with a prime symbol ('), are over-optimistic, since the compressor missed about 5 % of the triples. The column "best other" denotes the best, to our knowledge, compression ratio from other programs reported in the literature.

	No. triples	Inp. size (MB)	HDT +zip	HDT +7z	Best other	OFR +zip	OFR +7z
AEMET-1	1,018,815	139.2	0.55	0.41	0.8[c]	0.26	0.24
AEMET-2	2,788,429	517.8	0.70	0.30	1.1[b,c]	0.09	0.08
CN	137,484	18.8	0.95	0.80	0.78[d]	0.34	0.22
Events-Wikipedia	359,028	34.2	5.03	—	—	3.43	3.07
Jamendo	1,047,950	151.2	3.44	2.58	3.19[d]	2.58	2.02
LinkedMbd	6,147,996	891.6	1.88	1.50	1.01[a]	0.90	0.79
Mix	93,048	12.3	3.03'	2.65'	4.9[b]	2.18	1.93
Petrol	3,356,616	508.9	1.83	1.62	2.6[c]	1.86	1.62

[a] is from the k2-triples result, copied from [10]
[b] is from the algorithm RDSZ [9], as reported in [7]
[c] is from ERI-4k-Nodict [7], as reported in the same work
[d] is the JHD [12] result, with bzip2 backend compression

On the test machine, the average measured compression and decompression speeds were, respectively, 47 and 62 MB/s (including both OFR and second-stage processing), which can be considered as satisfactory for practical purposes. The compression speed ranged between 24 and 77 MB/s, with slowest compression measured for datasets containing large amounts of numbers, which is reasonable as they are handled in a sophisticated way, whereas decompression speed, ranging between 22 and 103 MB/s, seemed to depend primarily on the dataset size, with the smaller datasets being decompressed slower (presumably due to larger initialization overhead). A comparison of measured processing times of OFR and HDT is presented in Table 2. The OFR implementation is clearly faster, yet one should notice that the HDT implementation was written in Java, which may be somewhat slower than equivalent C++ code.

Table 2. Compression and decompression times (s). The test machine was running 64-bit Windows 7 and was equipped with an AMD Phenom II x6 3.0 GHz CPU and 8 GB of RAM.

	Compression time				Decompression time			
	HDT +zip	HDT +7z	OFR +zip	OFR +7z	HDT +zip	HDT +7z	OFR +zip	OFR +7z
AEMET-1	3.81	3.96	1.74	1.81	4.76	4.79	1.33	1.35
AEMET-2	14.87	16.95	7.28	7.71	15.72	15.79	5.99	5.93
CN	1.18	1.18	0.30	0.32	1.08	1.08	0.42	0.43
Events-Wikipedia	2.73	2.92	0.96	1.30	2.23	2.28	1.05	1.05
Jamendo	9.05	11.69	3.79	6.40	8.29	8.42	2.30	2.38
LinkedMbd	50.74	54.47	15.66	19.17	31.36	31.84	11.44	11.65
Mix	1.14	1.18	0.30	0.37	1.21	1.20	0.52	0.55
Petrol	16.34	23.77	8.32	12.26	23.10	23.47	7.47	7.77

5   Conclusion

We presented an RDF compression scheme surpassing the existing ones in compression ratio. This was achieved thanks to careful parsing of the RDF content, reordering it, distributing into multiple streams, and encoding each stream using, first, the most adequate specialized techniques, and then, an efficient general-purpose compression algorithm, such as LZMA. Our proof-of-concept implementation was also found to be relatively fast, regarding both compression and decompression. The future work will be to develop a query-supporting RDF compressed representation, based on the presented solution.

References

1. Abadi, D.J., Marcus, A., Madden, S., Hollenbach, K.J.: Scalable semantic web data management using vertical partitioning. In: Proceedings of the 33rd International Conference on Very Large Data Bases, pp. 411–422. ACM (2007)
2. Álvarez-García, S., Brisaboa, N.R., Fernández, J.D., Martínez-Prieto, M.A.: Compressed k2-triples for full-in-memory RDF engines. In: A Renaissance of Information Technology for Sustainability and Global Competitiveness. 17th Americas Conference on Information Systems. Association for Information Systems (2011)
3. Atre, M., Chaoji, V., Zaki, M.J., Hendler, J.A.: Matrix "bit" loaded: a scalable lightweight join query processor for RDF data. In: Proceedings of the 19th International Conference on World Wide Web, WWW 2010, Raleigh, North Carolina, USA, 26–30 April 2010, pp. 41–50. ACM (2010)

4. Brisaboa, N., Ladra, S., Navarro, G.: Compact representation of web graphs with extended functionality. Inf. Syst. **39**(1), 152–174 (2014)
5. Broekstra, J., Kampman, A., van Harmelen, F.: Sesame: a generic architecture for storing and querying RDF and RDF schema. In: Horrocks, I., Hendler, J. (eds.) ISWC 2002. LNCS, vol. 2342, pp. 54–68. Springer, Heidelberg (2002)
6. Cheng, L., Malik, A., Kotoulas, S., Ward, T.E., Theodoropoulos, G.: Scalable RDF data compression using X10. CoRR, abs/1403.2404 (2014)
7. Fernández, J.D., Llaves, A., Corcho, O.: Efficient RDF interchange (ERI) format for RDF data streams. In: Mika, P., Tudorache, T., Bernstein, A., Welty, C., Knoblock, C., Vrandečić, D., Groth, P., Noy, N., Janowicz, K., Goble, C. (eds.) ISWC 2014, Part II. LNCS, vol. 8797, pp. 244–259. Springer, Heidelberg (2014)
8. Fernández, J.D., Martínez-Prieto, M.A., Gutierrez, C.: Compact representation of large RDF data sets for publishing and exchange. In: Patel-Schneider, P.F., Pan, Y., Hitzler, P., Mika, P., Zhang, L., Pan, J.Z., Horrocks, I., Glimm, B. (eds.) ISWC 2010, Part I. LNCS, vol. 6496, pp. 193–208. Springer, Heidelberg (2010)
9. Fernández, N., Arias, J., Sánchez, L., Fuentes-Lorenzo, D., Corcho, Ó.: RDSZ: an approach for lossless RDF stream compression. In: Presutti, V., d'Amato, C., Gandon, F., d'Aquin, M., Staab, S., Tordai, A. (eds.) ESWC 2014. LNCS, vol. 8465, pp. 52–67. Springer, Heidelberg (2014)
10. Hernández-Illera, A., Martínez-Prieto, M.A., Fernández, J.D.: Serializing RDF in compressed space. In: Data Compression Conference (DCC) (2015)
11. Jiang, X., Zhang, X., Gao, F., Pu, C., Wang, P.: Graph compression strategies for instance-focused semantic mining. In: Qi, G., Tang, J., Du, J., Pan, J.Z., Yu, Y. (eds.) CSWS 2013. CCIS, vol. 406, pp. 50–61. Springer, Heidelberg (2013)
12. Joshi, A.K., Hitzler, P., Dong, G.: Logical linked data compression. In: Cimiano, P., Corcho, O., Presutti, V., Hollink, L., Rudolph, S. (eds.) ESWC 2013. LNCS, vol. 7882, pp. 170–184. Springer, Heidelberg (2013)
13. Maneth, S., Peternek, F.: A survey on methods and systems for graph compression. CoRR, abs/1504.00616 (2015)
14. Martínez-Prieto, M.A., Fernández, J.D., Cánovas, R.: Compression of RDF dictionaries. In: 27th ACM International Symposium on Applied Computing (SAC 2012) - Track The Semantic Web and Applications (SWA), pp. 1841–1848. ACM (2012)
15. Neumann, T., Weikum, G.: The RDF-3X engine for scalable management of RDF data. VLDB J. **19**(1), 91–113 (2010)
16. Pan, J.Z., Pérez, J.M.G., Ren, Y., Wu, H., Wang, H., Zhu, M.: Graph pattern based RDF data compression. In: Supnithi, T., Yamaguchi, T., Pan, J.Z., Wuwongse, V., Buranarach, M. (eds.) JIST 2014. LNCS, vol. 8943, pp. 239–256. Springer, Heidelberg (2015)
17. Urbani, J., Maassen, J., Drost, N., Seinstra, F.J., Bal, H.E.: Scalable RDF data compression with MapReduce. Concurrency Comput. Pract. Experience **25**(1), 24–39 (2013)
18. Weiss, C., Karras, P., Bernstein, A.: Hexastore: sextuple indexing for semantic web data management. PVLDB **1**(1), 1008–1019 (2008)
19. Wilkinson, K.: Jena property table implementation. In: SSWS (2006)

Reducing Large Semantic Graphs to Improve Semantic Relatedness

Teresa Costa[✉] and José Paulo Leal

Faculty of Sciences, CRACS and INESC-Porto LA,
University of Porto, Porto, Portugal
{teresa.costa,zp}@dcc.fc.up.pt

Abstract. In the previous research the authors developed a family of semantic measures that are adaptable to any semantic graph, being automatically tuned with a set of parameters. The research presented in this paper extends this approach by also tuning the graph. This graph reduction procedure starts with a disconnected graph and incrementally adds edge types, until the quality of the semantic measure cannot be further improved. The validation performed used the three most recent versions of WordNet and, in most cases, this approach improves the quality of the semantic measure.

Keywords: Semantic similarity · Linked data · Semantic graph

1 Introduction

This paper is part of an ongoing research [6, 12, 13] aiming at the development of a methodology for creating semantic measures taking as source any given semantic graph. This methodology, called *SemArachne*, does not require any particular knowledge of the semantic graph and is based on the notion of *proximity* rather than distance. It considers virtually all paths connecting two terms with weights depending on edge types. SemArachne automatically tunes these weights for a given semantic graph. The validation of this process was performed using WordNet 2.1 [8] with WordSimilarity 353 [9] data set with results better than those in the literature [13].

WordNet 2.1 has a smaller graph when compared with the recent versions of it or even other semantic sources, such as DBpedia or Freebase. Not only the number of nodes and edge types increases as the number of graph arcs expands enabling them to relate semantically a large number of terms, making graphs not only larger but also denser. Compute proximity in these conditions comes with a price. Since SemArachne considers all the paths, the number of paths to process tends to increase.

A rough measure of graph density is the maximum degree of all its nodes. However, consider it can be misleading since there may be a special node where all the edge types are applied. The real challenge is then the graph *average node degree*. SemArachne computes all paths connecting a pair of terms up tp a given

© Springer International Publishing Switzerland 2015
J.-L. Sierra-Rodríguez et al. (Eds.): SLATE 2015, CCIS 563, pp. 236–245, 2015.
DOI: 10.1007/978-3-319-27653-3_23

length. The node degree is the branching factor for the paths crossing that node. Hence, a high average node degree reduces the efficiency of the SemArachne measure.

The alternative explored in this paper to reduce graph density is to reduce the number of edge types while keeping all nodes, thus preserving the potential to relate a larger set of terms. The approach is to incrementally build a subgraph of the original semantic graph. This process starts with a full disconnected graph containing all the nodes. At each iteration, a new edge type is added until the semantic measure quality stops to improve. The result of this process is a subgraph where the semantic quality is maximized. The semantic measure used by SemArachne [12] had also some minor adjustments.

The rest of the paper is organized as follows. The next section surveys the state of the art on semantic relatedness. Section 3 summarizes previously published work and Sect. 4 details the approach followed to measure semantic relatedness is larger graphs. The experimental results and their analysis can be found in Sect. 5. Finally, Sect. 6 summarizes what was accomplished so far and identifies opportunities for further research.

2   Related Work

Semantic measures are widely used today to measure the strength of the semantic relationship between terms. This evaluation is based on the analysis of information describing the elements extracted from semantic sources.

There are two different types of semantic sources. The first one are unstructured and semi-structured texts, such as plain text or dictionaries. Texts have evidences of semantic relationships and it is possible to measure those relationships using simple assumptions regarding the distribution of words. This source type is mainly used by distributional approaches.

The second type of semantic sources is more general and includes a large range of computer understandable resources where the knowledge about elements is explicitly structured and modeled. Semantic measures based on this type of source rely on techniques to take advantage of semantic graphs or higher formal knowledge representations. This source type is mainly used by knowledge-based approaches.

Distributional approaches rely on the *distributional hypothesis* [11] that states that words in a similar context are surrounded by the same words and are likely to be semantically similar. There are several methods following this approach, such as the Spatial/Geometric methods [10], the Set-based methods [5], and the Probabilistic methods [7].

The knowledge-base approaches rely on any form of knowledge representation, namely semantic graphs, since they are structured data from which semantic relationships can be extracted. They consider the properties of the graph and elements are compared by analysing their interconnections and the semantics of those relationships. Several methods have been defined to compare elements in single and multiple knowledge bases, such as Structural methods [14,15,22,24], Feature-based methods [4,23,27] and Shannon's Information Theory methods [16,19–21].

Knowledge-based approaches have the advantage of controlling which edge types should be considered when comparing pairs of elements in the graph. They are also easier to implement than distributional methods and have a lower complexity. However they require a knowledge representation containing all the elements to compare. On the other hand, using large knowledge sources to compare elements is also an issue due of high computational complexity.

There are also hybrid approaches [2,3,18,24] that mix the knowledge-based and the distributional approaches. They take advantage of both texts and knowledge representations to estimate the semantic measure.

3   Previous Work

This section summarizes previously published work [12,13] that is the core of SemArachne and relevant for the graph reduction process described in the next section. The first subsection details on the semantic measure and the following subsection on the quality measure. The last subsection details on the fine tune process.

3.1   Semantic Measure

A semantic graph can be defined as $G = (V, E, T, W)$ where V is the set of nodes, E is the set of edges connecting the graph nodes, T is the set of edge types and W is a mapping of edge types to weight values. Each edge in E is a triplet (u, v, t) where $u, v \in V$ and $t \in T$.

The set W defines a mapping $w : T \mapsto \mathbb{Z}$. The bound of the absolute weight values[1] for all edge types is defined by

$$\Omega(G) \equiv max_{t_i \in T} \ | \, w(t_i) \, |$$

To measure the proximity between a pair of terms it is necessary to build a set of distinct paths that connects them by walking through the graph. A path p of size $n \in \mathbb{N}^+$ is a sequence of unrepeated nodes $u_0 \ldots u_n \forall_{0 \leq i,j \leq n} u_i \neq u_j$, linked by typed edges. It must have at least one edge and cannot have loops. A path p is denoted as follows:

$$p = u_0 \xrightarrow{t_1} u_1 \xrightarrow{t_2} u_2 \ldots u_{n-1} \xrightarrow{t_n} u_n$$

The weight of an edge depends on its type. The weight of a path p is the sum of weights of each edge, $\omega(p) = w(t_1) + w(t_2) + \ldots + w(t_n)$. The set of all paths of size n connecting the pair of concepts is defined as follows and its weight is the sum of all its sub paths.

$$P_{u,v}^n = \{u_0 \xrightarrow{t_1} u_1 \ldots u_{n-1} \xrightarrow{t_n} u_n : u = u_o \wedge v = u_n \wedge \forall_{0 \leq i,j \leq n} u_i \neq u_j\}$$

---

[1] This semantic measure accepts negative weights for some types of edges.

The semantic measure is based on the previous definition and also considers the path length. Δ is the degree of each node in each path. The proximity function r is defined by the following formula.

$$r(u,v) = \begin{cases} 1 & \leftarrow u = v \\ \frac{1}{\Omega(G)} \sum_{n=1}^{\infty} \frac{1}{2^n.n.\Delta(G)^n} \sum_{p \in P_{u,v}^n} \omega(p) & \leftarrow u \neq v \end{cases} \tag{1}$$

Given a graph with a set of nodes V, where $r : V \times V \mapsto [-1,1]$, the proximity function r takes a pair of terms and returns a "percentage" of proximity between them. The proximity of related terms must be close to 1 and the proximity of unrelated terms must be close to –1.

This definition of proximity depends on weights of transitions. The use of domain knowledge to define them has been proved a naïve approach since an "informed opinion" frequently has no evidence to support it and sometimes is plainly wrong. Also, applying this methodology to a large ontology with several domains can be hard. To be of practical use, the weights of a proximity based semantic relatedness measure must be automatically tuned. To achieve it, it is necessary to estimate the quality of a semantic measure for a given set of parameters.

3.2   Quality Measure

The purpose of the quality measure is to compute the quality of a semantic measure defined by (1) for a particular set of parameters. In order to simplify and optimize the quality measure, it is necessary to factor out weights from the semantic measure definition. Thus its quality may be defined as function of a set of weight assignment.

The first step is to express the semantic measure in terms of weights of *edge types*. Consider the set of all edge types T with $\sharp T = m$ and the weight of its elements $w(t), \forall t \in T$. The second branch of (1) can be rewritten as follows, where $c_i(a,b), i \in \{1..m\}$ are the coefficients of each edge type.

$$r(a,b) = \alpha \sum_{n_1}^{\infty} \beta \sum_{P_j \in \mathbb{P}} \sum_{t \in P_j} w(t) = \sum_{i=1}^{m} c_i(a,b) \cdot w(t_i)$$

Edge type weights are independent of the arguments of r but the coefficients that are factored out depend of these arguments. It is possible to represent both the weights of edges and their coefficients, $(w(t_1), w(t_2), \ldots, w(t_k)) = \boldsymbol{w}$ and $(c_1(a,b), c_2(a,b), \ldots, c_m(a,b) = \boldsymbol{c}(a,b))$ respectively, by defining a standard order on the elements of T. This way the previous definition of r may take as parameter the weight vector, as follows

$$\boldsymbol{w}(a,b) = \boldsymbol{c}(a,b) \cdot \boldsymbol{w}$$

The method commonly used to estimate the quality of a semantic relatedness algorithm is to compare it with a benchmark data set containing pairs of words and their relatedness. The *Spearman's rank order correlation* is widely used to make this comparison.

Consider a benchmark data set with the pairs of words (a_i, b_i) for $1 \leq i \leq k$, with a proximity x_i. Given the relatedness function $r_w : S \times S \mapsto \Re$ let us define $y_i = r_w(a_i, b_i)$. In order to use the Spearman's rank order coefficient both x_i and y_i must be converted to the ranks x_i' and y_i'.

The Spearman's rank order coefficient is defined in terms of x_i and y_i, where x_i are constants from the benchmark data set. To use this coefficient as a quality measure it must be expressed as a function of w. Considering that $y = (r_w(a_i, b_i), \ldots, r_w(a_n, b_n))$ then $y = Cw$, where matrix C is a $n \times m$ matrix and where each line contains the coefficients for a pair of concepts and each column contains coefficients of a single edge type. Vector w is a $m \times 1$ matrix with the weights assigned to each edge type. The product of these matrices is the relatedness measure of a set of concept pairs.

Considering $\rho(x, y)$ as the Spearman's rank order of x and y, the quality function $q : \Re^n \mapsto \Re$ using the benchmark data set x can be defined as

$$q_x(w) = \rho(x, Cw) \tag{2}$$

The next step in the SemArachne methodology is to determine a w that maximizes this quality function.

3.3 Fine Tuning Process

Genetic algorithms are a family of computational models that mimic the process of natural selection in the evolution of species. This type of algorithms uses concepts of *variation, differential reproduction* and *heredity* to guide the co-evolution of a set of problem solutions. This algorithm family is frequently used to improve solutions of optimization problems [29].

In the SemArachne the candidate solution – *individual* – is a weight values vector. Consider a sequence of weights (the genes), $w(t_1), w(t_2), \ldots, w(t_k)$, taking integer values in a certain range, in a standard order of edge types. Two possible solutions are the vectors $v = (v_1, v_2, \ldots, v_k)$ and $t = (t_1, t_2, \ldots, tn)$. Using crossover, it is easy to recombine the "genes" of both "parents" resulting in $u = (v_1, t_2, \ldots, t_{n-1}, v_k)$.

This is a closer representation of the domain than the typical binary one. It can also be processed more efficiently with large number of weights. In this tuning process the genetic algorithm only have a single kind of mutation: randomly selecting a new value for a given "gene".

The fitness function plays a decisive role in the selection of the new generation of individuals. In this case, individuals are the vector of weight values w, hence the fitness function is in fact the quality function previously defined in (2).

4   Graph Reduction Procedure

The previous section explained how to tune the weights of a semantic measure by using a genetic algorithm with an appropriate quality function. This section

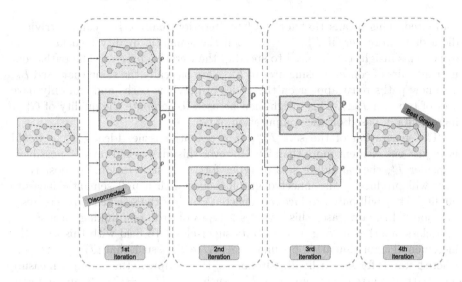

Fig. 1. Semantic graph reduction procedure

introduces a procedure for selecting a subgraph of the original semantic source with a reduced density by repeatedly applying that procedure.

Figure 1 depicts the overall strategy. It starts with a fully disconnected graph by omitting all the edges. The small graph on the left in Fig. 1 shows the arcs as dotted lines to denote the original connections. When a single property (edge type) is added to this graph a number of paths is created. If the original graph has n property types then one can create n different subgraphs. The quality of these graphs can be measured using the approach described in the last section. The best of these candidates is the selected graph for the first iteration. This process continues until the quality of the candidate graphs cannot be further improved.

More formally, consider a semantic graph $G = (V, E, T, W)$ where V is the set of nodes, E is the set of edges connecting the graph nodes, T is the set of edge types and W is a mapping of edge types to weight values. The initial graph of this incremental algorithm is $G_0 = (V, \emptyset, \emptyset, \emptyset)$. This is a totally disconnected graph just containing the nodes from the original graph, i.e., edges, types and weights are all the empty set.

Each iteration builds a new graph $G_{k+1} = (V, E_{k+1}, T_{k+1}, W_{k+1})$ based on $G_k = (V, E_k, T_k, W_K)$. The new set of types T_{k+1} has all the types in T_k. In fact, several candidate G_k^i can be considered, depending on the types in $T - T_k$ that are added to T_{k+1}. The arcs of E_{k+1}^i are those in E whose type is in T_{k+1}^i. The general idea is to select the G_{k+1}^i that produces an higher increment on semantic measure quality. This algorithm stops when no candidate is able to improve it.

In general, computing the semantic measure quality of G_{k+1}^i is a time consuming task. However, there are some ways to make it more efficient. As shown in Fig. 1, if G_{k+1}^i is not a connected graph then the quality measure cannot be computed. This means that for the first iteration many G_{k+1}^1 can be trivially discarded. Moreover, if $E_{k+1}^i = E_k^i$ then the semantic quality measure is the same. This insight can be used to speedup the iterative process. The paths connecting pairs of concepts using arcs in E_{k+1} are basically the same that used E_k. The new paths must appear on the nodes of previous paths and can only have arcs of types in T_{k+1}. This insight can be used to compute the quality of G_{k+1}^i incrementally based on the computation of G_k^i.

The generation of the sets T_{k+1}^i is a potential issue. Ideally T_{k+1}^i would have just one element more than T_k^i. However this may not always be possible[2]. Consider T_1^i, the candidate sets of types for the first iteration. In most cases they will produce a disconnected graph, hence with a null semantic measure quality. They will only produce a connected graph if the selected type creates a taxonomy. In many cases this involves 2 types of arcs: one linking an instance to a class, another linking a class to its super-class. To deal with this issue the incremental algorithm attempts first to generate T_{k+1}^i such that $\sharp T_{k+1}^i = \sharp T_k^i + 1$, where $\sharp$ stands for set cardinality. In none of these improve the semantic measure quality then it attempts to generate T_{k+1}^i such that $\sharp T_{k+1}^i = \sharp T_k^i + 2$, and so forth.

5   Validation

The validation of SemArachne was performed using the semantic graphs of different versions of WordNet along with three different data sets.

WordNet[3] [8] is a widely used lexical knowledge base of English words. It groups nouns, verbs, adjectives and adverbs into *synsets* – a set of cognitive synonyms – that expresses distinct concepts. These *synsets* are linked by conceptual and lexical relationships. The validation process used three different data sets: WordSimilarity-353[4] [9] Rubenstein & Goodenough [25] (RG65) and Miller & Charles [17] (MC30).

Table 1 compares the performance of SemArachne against the state of the art for methods using the same knowledge-based approach. For WordNet 2.1, SemArachne achieves a better result than those in the literature when using WordSim-353 data set. Using WordNet 3.1 as semantic graph, SemArachne produces also a better semantic quality than those in the literature. Although results are not the best in the WordNet 3.0, despite the data set used, they have the same order of magnitude.

The quality of the semantic measure produced with graph reduction was validated against several approaches in the literature. An advantage of this methodology is the ability of measure the semantic relatedness regardless the semantic graph used and produce comparable results for each semantic graph and data set. It is also scalable, since it handles gradually larger graphs.

---

[2] However, so far this situation has not yet occurred in validation.

[3] http://wordnet.princeton.edu/.

[4] http://www.cs.technion.ac.il/~gabr/resources/data/wordsim353/wordsim353.html.

Table 1. Spearman correlation of SemArachne compared with literature

Graph	Data set	Edges selected	SemArachne correlation	Literature correlation	Author
WordNet 2.1 (26 edge types)	MC30	14	0.81	**0.82**	Strube et al. [28] 2006
	RG65	8	0.60	**0.86**	
	WS-353	21	**0.45**	0.36	
WordNet 3.0 (47 edge types)	MC30	16	**0.80**	N/A	Agirre et al. [1] 2009
	RG65	9	0.63	**0.78**	
	WS-353	20	0.48	**0.56**	
WordNet 3.1 (64 edge types)	MC30	14	**0.97**	0.87	Siblini et al. [26] 2013
	RG65	8	**0.94**	0.92	
	WS-353	24	**0.54**	0.50	

6 Conclusion

As semantic graphs evolve they become larger. Since larger graphs relate more terms this improves their potential as semantic sources for relatedness measures. However, these larger graphs are also a challenge, in particular to semantic measures that consider virtually all paths connecting two nodes, as is the case of SemArachne.

The major contribution of this paper is an incremental approach to select a subgraph with a reduced number of edge types (arcs) but with the same number of entities (nodes). This approach starts with a totally disconnected graph, at each iteration adds an arc type that increases the quality of the semantic measure, and stops when no improvement is possible.

These contributions were validated with different versions of WordNet, a medium size graph typically used as semantic source for relatedness measures. Although this is not the kind of large semantic graphs to which this approach is targeted, it is convenient for initial tests due to its relatively small size.

In the WordNet graph the reduction of properties is not so expressive, since the total number of properties is comparatively small. The obtained subgraphs do not always improve the quality of the SemArachne measure, but produce a result that is similar, and in most cases better, than best method described in the literature for that particular graph.

The immediate objective of the SemArachne project is to extend the validation presented in this paper to other data sets and, most of all, to other graphs. Massive graphs with very high density, such as Freebase, are bound to create new and interesting challenges. Another important consequence of this graph reduction procedure is that it decouples the original graph from the actual semantic source. Thus SemArachne can be extended to process multiple semantic graphs (with shared labels) and create an unified semantic measure combining their semantic power.

Acknowledgments. Project "NORTE-07-0124-FEDER-000059" is financed by the North Portugal Regional Operational Programme (ON.2 - O Novo Norte), under the National Strategic Reference Framework (NSRF), through the European Regional Development Fund (ERDF), and by national funds, through the Portuguese funding agency, Fundação para a Ciência e a Tecnologia (FCT).

References

1. Agirre, E., Alfonseca, E., Hall, K., Kravalova, J., Paşca, M., Soroa, A.: A study on similarity and relatedness using distributional and wordnet-based approaches. In: Proceedings of Human Language Technologies: The 2009 Annual Conference of the North American Chapter of the Association for Computational Linguistics, pp. 19–27. Association for Computational Linguistics (2009)
2. Banerjee, S., Pedersen, T.: An adapted lesk algorithm for word sense disambiguation using wordnet. In: Gelbukh, A. (ed.) CICLing 2002. LNCS, vol. 2276, pp. 136–145. Springer, Heidelberg (2002)
3. Banerjee, S., Pedersen, T.: Extended gloss overlaps as a measure of semantic relatedness. IJCAI **3**, 805–810 (2003)
4. Bodenreider, O., Aubry, M., Burgun, A.: Non-lexical approaches to identifying associative relations in the gene ontology. In: Pacific Symposium on Biocomputing, p. 91. NIH Public Access (2005)
5. Bollegala, D., Matsuo, Y., Ishizuka, M.: Measuring semantic similarity between words using web search engines. In: Proceedings of the 16th International Conference on World Wide Web 2007, pp. 757–766 (2007)
6. Costa, T., Leal, J.P.: Challenges in computing semantic relatedness for large semantic graphs. In: Proceedings of the 18th International Database Engineering and Applications Symposium, pp. 376–377. ACM (2014)
7. Dagan, I., Lee, L., Pereira, F.C.: Similarity-based models of word cooccurrence probabilities. Mach. Learn. **34**(1–3), 43–69 (1999)
8. Fellbaum, C.: WordNet. Wiley Online Library, New York (1999)
9. Gabrilovich, E.: The WordSimilarity-353 test collection. http://www.cs.technion.ac.il/gabr/resources/data/wordsim353/
10. Ganesan, P., Garcia-Molina, H., Widom, J.: Exploiting hierarchical domain structure to compute similarity. ACM Trans. Inf. Syst. (TOIS) **21**(1), 64–93 (2003)
11. Harris, Z.S.: Distributional structure. In: Hiż, H. (ed.) Papers on syntax, pp. 3–22. Springer, The Netherlands (1981)
12. Leal, J.P.: Using proximity to compute semantic relatedness in RDF graphs. Comput. Sci. Inf. Syst. **10**(4), 1727–1746 (2013)
13. Leal, J.P., Costa, T.: Multiscale parameter tuning of a semantic relatedness algorithm. In: 3rd Symposium on Languages, Applications and Technologies, SLATE, pp. 201–213 (2014)
14. Li, Y., Bandar, Z.A., McLean, D.: An approach for measuring semantic similarity between words using multiple information sources. IEEE Trans. Knowl. Data Eng. **15**(4), 871–882 (2003)
15. Li, Y., McLean, D., Bandar, Z.A., O'shea, J.D., Crockett, K.: Sentence similarity based on semantic nets and corpus statistics. IEEE Trans. Knowl. Data Eng. **18**(8), 1138–1150 (2006)
16. Lin, D.: An information-theoretic definition of similarity. ICML **98**, 296–304 (1998)
17. Miller, G.A., Charles, W.G.: Contextual correlates of semantic similarity. Lang. Cogn. Process. **6**(1), 1–28 (1991)

18. Patwardhan, S., Banerjee, S., Pedersen, T.: Using measures of semantic relatedness for word sense disambiguation. In: Gelbukh, A. (ed.) CICLing 2003. LNCS, vol. 2588, pp. 241–257. Springer, Heidelberg (2003)
19. Pirró, G.: A semantic similarity metric combining features and intrinsic information content. Data Knowl. Eng. **68**(11), 1289–1308 (2009)
20. Pirró, G., Euzenat, J.: A feature and information theoretic framework for semantic similarity and relatedness. In: Patel-Schneider, P.F., et al. (eds.) ISWC 2010, Part I. LNCS, vol. 6496, pp. 615–630. Springer, Heidelberg (2010)
21. Pirró, G., Seco, N.: Design, implementation and evaluation of a new semantic similarity metric combining features and intrinsic information content. In: Meersman, R., Tari, Z. (eds.) OTM 2008, Part II. LNCS, vol. 5332, pp. 1271–1288. Springer, Heidelberg (2008)
22. Rada, R., Mili, H., Bicknell, E., Blettner, M.: Development and application of a metric on semantic nets. IEEE Trans. Syst. Man Cybern. **19**(1), 17–30 (1989)
23. Ranwez, S., Ranwez, V., Villerd, J., Crampes, M.: Ontological distance measures for information visualisation on conceptual maps. In: Meersman, R., Tari, Z., Herrero, P. (eds.) OTM 2006 Workshops. LNCS, vol. 4278, pp. 1050–1061. Springer, Heidelberg (2006)
24. Resnik, P.: Using information content to evaluate semantic similarity in a taxonomy. In: IJCAI, pp. 448–453 (1995)
25. Rubenstein, H., Goodenough, J.B.: Contextual correlates of synonymy. Commun. ACM **8**(10), 627–633 (1965)
26. Siblini, R., Kosseim, L.: Using a weighted semantic network for lexical semantic relatedness. In: RANLP, pp. 610–618 (2013)
27. Stojanovic, N., Maedche, A., Staab, S., Studer, R., Sure, Y.: SEAL: a framework for developing SEmantic portALs. In: Proceedings of the 1st International Conference on Knowledge Capture, pp. 155–162. ACM (2001)
28. Strube, M., Ponzetto, S.P.: Wikirelate! Computing semantic relatedness using wikipedia. AAAI **6**, 1419–1424 (2006)
29. Whitley, D.: A genetic algorithm tutorial. Stat. Comput. **4**(2), 65–85 (1994)

A Mixed Approach for the Representation of Nutritional Information Through XML-to-OWL Mappings

Vanesa Espín$^{(\boxtimes)}$, Manuel Noguera, and María V. Hurtado

Departamento de Lenguajes y Sistemas Informáticos, University of Granada,
E.T.S.I.I.T., c/Daniel Saucedo Aranda s/n, 18071 Granada, Spain
{vespin,mnoguera,mhurtado}@ugr.es

Abstract. Semantic Web technologies (SWTs), such as XML and OWL ontologies are increasingly being used to represent information in different domains. However, these capabilities are not indistinctly provided by each SWT. XML, although not being considered a SWT in itself, stands at the syntactic level of the Semantic Web stack, and is more suitable for efficient information structure and retrieval in interactive software applications. OWL language, on the other hand, is more suitable for background reasoning and consistency checking purposes. In this paper, we introduce a mixed approach for the information representation and knowledge sharing in the nutritional domain, aiming to explode XML and OWL benefits. This approach is included in NutElCare (a nutritional recommender system). In it, diets are represented through XML documents contained in an XML repository and the knowledge base is composed of several OWL ontologies which interact to provide recommendations. In this design, XSLT transformations play an important role, allowing the mappings from XML diets to the OWL ontologies, so that elevating the syntactic representation of the XML documents to the semantic level of OWL ontologies. Altogether, they configure a system architecture that keeps the system timely responsive through a seamless linkage between XML and OWL representations.

Keywords: OWL ontology · Ontology enrichment · Ontology population · Ontology reasoning · Semantic recommender systems · XML · XSLT

1 Introduction

Several reasons stimulate the rapid increase of the application of Semantic Web Technologies (SWTs) in the representation of information and knowledge in different software related fields in the last years. Some of these motivations are the abilities of reasoning to extract meaningful conclusions from encoded knowledge and the exchange, linkage and reuse of this knowledge from different systems, processes and applications [14]. XML (EXtensible Markup Language) was designed for describing data in the World Wide Web. Although it is not considered a SWT in itself, it stands at the syntactic level of the Semantic Web stack. It is platform and software independent and allows the representation, storage and exchange of data when a common syntax has

J.-L. Sierra-Rodríguez et al. (Eds.): SLATE 2015, CCIS 563, pp. 246–257, 2015.
DOI: 10.1007/978-3-319-27653-3_24

been agreed. XML it is not suitable for consistency checking or conceptual interrelationship from a semantic standpoint, even in the same domain of knowledge. OWL (Web Ontology Language), currently in OWL 2 version [15], is a formal language based on Description Logics [1]. In contrast to XML, it supports the representation of the domain knowledge through classes, properties and instances to be used in a distributed environment such as the World Wide Web [3]. Ontologies are one of the main components of the Semantic Web. They provide universal semantics, easy knowledge sharing and unambiguous interpretation of concepts by means of formal model-theoretic semantics. Ontologies represented in OWL can make use of the automated reasoning capabilities that Description Logics provide, allowing the support of this reasoning to infer new knowledge. However, the use of an OWL ontology to represent a big amount of structured data in which only a small part of this information is demanded at a required moment for inference purposes, could result in inefficient reasoning. In this case, a repository of XML documents for the efficient management, storage and representation of information can be designed, and instead retrieve and bind this information on-demand, only when it is required. Ontologies are also used in information retrieval for indexing documents, providing a semantic classification for the information of the documents [11]. Nevertheless, the process of linkage between XML documents or repositories and OWL ontologies needs some operations, such as enrichment and population of the ontology. Ontology enrichment is the task of extending an existing ontology with additional concepts and semantic relations and placing them at the correct position in the ontology [17]. It can be considered a sub-discipline of ontology learning and its application is typically motivated by one of the two following goals. One goal consists in the discovery of new knowledge through the analysis of an existing ontology or knowledge base, applying automated inference techniques. The axioms of the inferred knowledge are added to the ontology. The other motivation is the completion of the ontology with new information about the same domain [4]. Ontology population is the task of adding new instances or individuals to the ontology, which can be later *unpopulated*, i.e. removed from the ontology. These operations need the use of mediator technologies for its procedure, such as XSL (EXtensible Stylesheet Language) transformations, XSLT, which are able to transform XML documents into other formats.

NutElCare (Nutrition for Elder Care) [7] is a recommender system which allows elderly people to set up their own healthy diet plans according their needs due to aging and considering their food preferences, as well as possible allergies or contraindications and previous ingestions, i.e., what they have eaten in the past. In the system, diets are represented as XML documents and stored in an XML server repository. These documents are classified using indexes as instances of concepts in an ontology for being retrieved only when they are required for reasoning in the recommendation processes.

In this paper we introduce the design of the information, knowledge and software architectures of NutElCare focusing in the XML diets representation, and the enrichment of the ontology to manage the model contained in the XML Schema. We explain as well, the processes of information retrieval using on-demand binding and population which allow the knowledge base to make more efficient reasoning over the contents of the diets. In our approach, these processes are based in XSLT transformations from the syntactic level of the XML diet model definition to a semantic level supplied by an OWL diet model definition.

The remaining of this paper is organised as follows. Section 2 introduces some work related to ontology enrichment and population. In Sect. 3, we explain the different NutElCare recommendations and the ontologies contained in the system's knowledge base, outlining the techniques for diets classification and indexation in the nutritional ontology. In Sect. 4, the representation of diets and the processes of ontology enrichment and population are explained. Finally, in Sect. 5, the conclusions of our work are presented.

2 Related Work

Nowadays, researchers are still working in the automatic ontologies enrichment from different information sources. Often the ontology enrichment operations need some previous preparation of the data by hand, and even, some adjustments after the process. Thus, most of ontology enrichments are semi-automatic processes where efforts are focused in the minimization of hand working. One widespread approach is the use of generic XSL style sheets to perform XSLT transformations from the source information representation language to the target ontology language, generally OWL. Other approaches define their own XSL files for performing the transformations. This is the case of [2], in which XML patent documents are converted into OWL by means of XSL files and an XSLT processor. When the information to be imported for enriching an ontology is formatted into XML, the elements of the corresponding XML Schema, contained in an XSD document, are typically used to create the linkage between them. This connection is established through the definition of mapping rules from the structure and elements of the XML Schema to the OWL vocabulary. Some methods to generate OWL model documents from XSD documents have been presented in several works. In [9], authors propose a method based in a set of predefined mapping rules between XML Schema and OWL, and supply a Java toolkit that implements the mapping process. A similar approach is introduced in [3], but with different mapping rules; in this case, the authors provide an online tool, XML2OWL-XSLT, for transforming uploaded XSD documents. X2OWL tool [10], improves previous works addressing complex cases in mapping processes that arise from the reuse of global types and elements. However, this approach is based in the generation of an OWL ontology from scratch and does not deal with the mapping to an existing ontology, neither with references and imports to external ontologies in the Schema. A very good survey of the current tools that support the generation of OWL representations from XML documents with enrichment purposes is presented in [12]. In our work, the transformations from XML diet documents to OWL language use the XML Schema documents to keep the syntactic model of the diets in the system and XML validation purposes. Although the use of generic available online tools for these transformations, such as [3, 9, 19], seemed promising, we decided the definition of our own XSLT transformations and mapping rules, since some problems raised when trying these tools with our XML Schema, such as different OWL syntax, version or specie than expected. In other cases, those requirements were met, but the generated document needed some hand arrangements to be opened in the ontology editor, or to fit the expected model.

Other problems were the handling of annotations or namespaces from different ontologies importation. Finally, in some cases the mapping rules had to be redefined.

In contrast to the enrichment process, the population of an ontology can be fully achieved automatically. However, when the instances to populate the ontology come from heterogeneous sources, this process may become more complex and require different techniques to mediate, such as instance matching, validating and grouping [5, 17]. In our work, the automatic population of a retrieved XML diet from the repository is possible once the enrichment process has been performed using the generated indexes classified in the ontology as instances. A simple XSL style sheet binds each instanced element to its correspondent added concept in the ontology. Unlike other strategies for retrieving information in the population process, in our approach, this process is only carried out when an instantiation of one diet from the repository is required and unpopulated when it is no longer needed for the system operation, enhancing responsiveness and interactivity in the recommender system.

3   NutElCare

NutElCare is a semantic nutritional recommender system to provide healthy diet plans to the elderly based on their nutritional needs and preferences. In this section, we describe the different types of recommendations carried out by the system and the ontologies that comprise the knowledge base. We also present the classification and indexation of diets in the nutritional ontology.

3.1   Nutritional Recommendations

Recommendations in NutElCare are based on two different techniques:

- Knowledge-based techniques, which use knowledge about users and items to generate a recommendation by reasoning about what items meet the user's requirements. These techniques are used to obtain a healthy diet to fit the nutritional requirements identified from the user profile.
- Content-based techniques, where the recommendation process consists in learning from the user's alimentary behaviour and recommending items that are similar to their top rated meals or dishes. The content-based recommendation allows the users to make variations on the selected diet to fit their taste preferences, or availability of ingredients taking into account their allergenic contraindications and what food has already been ingested during the week, offering alternatives to the original diet plan based on these factors. These recommendations are always nutrient guided, providing alternative suggestions of similar conditions, to continue meeting the initial healthy requirements of the diet. The system learns from the user selections to improve further recommendations with user inferred preferences.

3.2 Knowledge Base

The knowledge base of NutElCare is represented as an OWL ontology resulting from the merge operation of three different ontologies: NUSPro, Food Ontology and Nutritional Ontology. The manual edition and visualization of the ontologies is carried out in Protégé[1] and the management of the ontologies performed by the NutElCare system is made through the OWL-API Java library [16], and the reasoner used is Pellet [18]. Next, we briefly describe the aforementioned ontologies and their role in the knowledge base in order to provide recommendations.

– The Nutritional User Profile Ontology (NUSPro) has been designed as an extension of GUMO (General User Model Ontology) [13] for representing users in nutritional domains.
– The Food Ontology has been obtained from the *Food Products* branch of the Agrovoc FAO Thesaurus [6] of the United Nations. It has been extended with nutritional properties of food, new food classifications and new food instances.
– The Nutritional Ontology establishes the concepts related to nutritional restrictions and requirements for user profiles. The central concept of the ontology is the *Diet* class whose descendant concepts –or subclasses– are used to classify the different existing diets in the system. This classification has been formerly agreed by the nutritional experts that supervise this project. When a new diet is being introduced in the system through the user interface, it must be classified according to the taxonomy already established in the ontology. This can be achieved through a simple XSLT transformation, using the Java XML library, from the *Diets* OWL Class to a form JSP page, which users fill for the classification of the diet. Once the form is filled an URL with a new Id of diet is generated and inserted into the ontology as a subclass of the terms of the classification selected by the user. This Id is an index from the ontology to the document in XML repository. Figure 1 outlines this process. In it, the user classifies the diet as *elderly diet* and *vegan diet*, so, when the new instance is created, a new index is generated, Diet25, and it is added to the OWL ontology as:

```
<owl:NamedIndividual rdf:about="&nutelcare;Diet25">
        <rdf:type rdf:resource="&nutelcare;elderly_diet"/>
        <rdf:type rdf:resource="&nutelcare;vegan_diet"/>
</owl:NamedIndividual>
```

Having the diets indexed in the nutritional ontology, the system is able to perform a knowledge-based recommendation of one diet which fits the user profile, but it is not able to personalize it with the preferences of the specific user, i.e., it cannot achieve the content-based recommendation required to allow the selection of alternatives over the food items contained in the diet. Consequently, the representation of the diets and the corresponding components must be likewise handled by the nutritional ontology.

The next section explains the representation of diets in the NutElCare system architecture and how this representation provides support to the content-based recommendation reasoning.

---

[1] Protégé, OWL Ontology Editor. http://protege.stanford.edu.

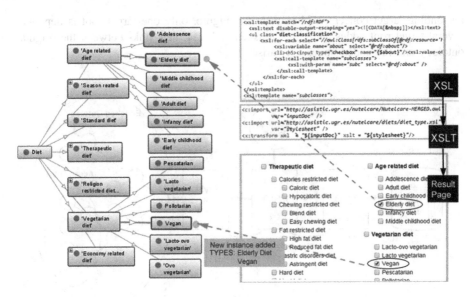

Fig. 1. Diets classification and indexation from the XML diets repository through XLST.

4   Diets Model Design and Information Retrieval

In order to perform recommendations over the contents of one diet, those contents must be represented in the ontology in the form of concepts, relations and individuals. The representation of a diet comprises the representation of each daily ingestion which in turn contains from three to five intakes (at least breakfast, lunch and dinner), and each intake holds several dishes with several meals, food ingredients and preparations. However, maintaining all this information over time in the ontology, for being used only a small part of this time, can affect to the system efficiency. Hence, the data representation of diets was carried out through XML documents based on a previously agreed syntactic structure and terminology defined in an XML Schema, which constitutes the syntactic model of the diets located in the XML repository.

4.1   Ontology Enrichment

The first step in order to allow the system reasoning over the contents of one diet is ontology enrichment for transferring the diet model of the XML Schema into the nutritional ontology. In this way, we build a semantic model of the diet from the syntactic one. The ontology enrichment process followed in our approach is depicted in Fig. 2. In it, an XSD document with the syntactic model of the XML diets is obtained. Next, the mapping rules between XSD and OWL have to be established and applied through XSLT transformations, generating an OWL model of the diet. Finally, the nutritional ontology is enriched by adding the generated diet OWL model at the corresponding target concept. Although in the Figure, only one target root is depicted, several target root nodes can be established and also, new links between the original ontology and the added nodes may arise.

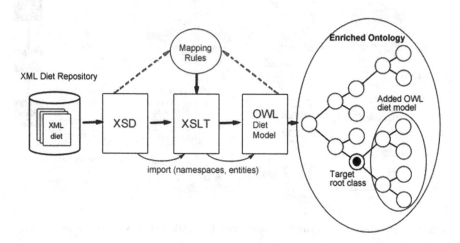

Fig. 2. Ontology enrichment process from the XML repository of diets.

The mapping between the XML Schema nodes and the OWL concepts is established through an XSLT transformation using an XSL style sheet following the rules summarized in Table 1. Note that these rules are defined only for our diet model and some of them differ from the defined in other works.

Table 1. Mapping Rules between our XSD diet model to OWL diet model.

XSD Nodes	OWL Concepts
xsd:element, with nested elements or at least one attribute	owl:Class + owl:ObjectProperty
other xsd:element	owl:ObjectProperty
named xs:complexType	owl:Class
named xs:simpleType	owl:DatatypeProperty
xs:minOccurs, xs:maxOccurs	owl:minCardinality, owl:maxCardinality
xs:choice	owl:unionOf
xs:sequence	owl:intersectionOf

The annotation of diets to be stored in the repository is made through the NutElCare user interface using a different XSL style sheet and parsed with the XML Schema automatically, which minimize the introduction of annotation errors.

A graphical representation of the XSL Schema model of diets is displayed in Fig. 3. It is important to point out that in the figure, FoodItem references an individual

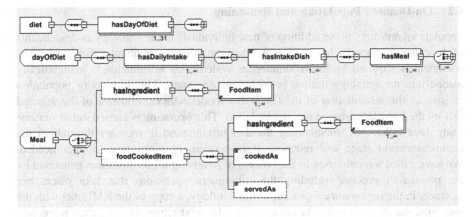

Fig. 3. Graphical model of a diet from its XML Schema generated with XSDDiagram (http://regis.cosnier.free.fr/?page=XSDDiagram).

from the food ontology and it is linked to the ontology trough the *attribute* ref="&nutelcare:FoodItem" and its corresponding namespace. This way, the food items of the diets are linked to the food ontology concepts.

An example of the transformation is the result of the application of the mapping rules over the Meal node obtaining the following OWL Meal model is shown next.

```
<owl:Class rdf:about="Meal">
<owl:equivalentClass>
 <owl:Class>
   <owl:unionOf rdf:parseType="Collection">
    <owl:Restriction>
      <owl:onProperty rdf:resource="hasIngredient"/>
      <owl:someValuesFrom rdf:resource="&nutelcare;Foods"/>
    </owl:Restriction>
    <owl:Restriction>
      <owl:onProperty rdf:resource="hasFoodCookedItem"/>
      <owl:someValuesFrom rdf:resource="foodCookedItem"/>
    </owl:Restriction>
   </owl:unionOf>
 </owl:Class>
</owl:equivalentClass>
</owl:Class>
```

Once the ontology is enriched with the new OWL diet model concepts and relations, it is able to support reasoning with the contents of an XML diet. In the next section, the ontology population and subsequent reasoning processes are introduced.

4.2   On-Demand Population and Reasoning

Populate an ontology is the addition of new individuals to the ontology as descendants of the existing concepts. In our system, when the information about a diet in an XML document is required by the recommender system, this information is automatically loaded into the ontology until it is no longer necessary. This process of population consists in the instantiation of the OWL diet model with the contents of the selected diet by the knowledge-based recommendation. This operation is carried out at runtime using Java OWL-API, maintaining the diet instantiation in memory throughout the recommendation stage and releasing it from memory when is no longer needed. We have called *unpopulation* to the process of removing all the instances generated by the population process including the subsequent operations that take place. For instance, in the *unpopulation* process of the ontology, a copy of the XML diet with the user food variations is stored in the repository for next retrievals. This copy represents the current diet for the user, which is different from the original one, because it is personalized with user special requirements and preferences. This is the diet used in the next population process for the recommendations to the same user. Also, a log-copy is generated and stored weekly, containing the ingested food in the week, for the availability of the historical nutritional records for each user.

Reasoning over the ontology allows the system to offer nutrient-guided variations over the food that the instantiated diet contains, in order to personalize diets adjusting to user needs and preferences. This is achieved through semantic similarity metrics with the individuals of the food ontology and the application of the general and nutritional restrictions taking into account the food already ingested by the user in the same week. The process of reasoning over the food ontology using semantic similarity in NutElCare is explained in [8]. For instance, if a user decides changing the meal "Grilled Salmon", the system calculate the possible variations and offers a healthy list of alternatives. This calculation is computed on the basis of the just-mentioned similarity. Let's assume that the user selects "Grilled Tuna" instead. Then, the system checks the future ingestions of the diet performing new reasoning to check whether it needs to adjust the diet plan for the remaining days of the week, to keep meeting the nutritional requirements. For instance, our system only allows recommending the same fish a maximum of 2 times a week. Suppose the user has already consumed *tuna* on Monday and in the diet *tuna* appears again in Saturday. But today is Thursday, and the user decides to change the "Grilled Salmon" by "Grilled Tuna". At this time, the system checks the rest of the diet detecting an ontology inconsistency of type "repeated food." In order to resolve this inconsistency, the system must perform a calculation of semantic similarity for substituting the *tuna* meal for Saturday by a different one but nutritionally similar in which *tuna* is not contained.

Also, reasoning over the ontology with the instantiation of diets allows many other knowledge inferences, for instance, for monitoring purposes. One example is the calculation of the daily nutritional properties, as the total consumed calories in one day, checking also whether this value fits the user nutritional daily requirements and sending notifications if it detects lacks or excesses.

5   Conclusions and Future Work

OWL and XML can work together to further exploit the each other benefits and overcome their weaknesses. The use of an OWL ontology to represent a big amount of structured data in which only a small part of this information is demanded at a required moment can lead to inefficient responsiveness regarding to end users. In order to lighten the ontology of unnecessary data in the reasoning process, a repository of XML documents can be used for storing this information and retrieve a document on-demand, only when it is required.

NutElCare is a semantic nutritional recommender system whose recommendations are accomplished through several reasoning processes over the ontologies of its knowledge base. For the purpose of keeping efficiency in the involved reasonings, an XML diets repository is used for storing the diets information. When a new diet is incorporated to the system through the system's user interface, it is automatically annotated in an XML document and classified in the ontology for indexing in further retrievals. In this work, we have explained the information representation, indexation and retrieval of XML diets in our recommender system for being used in nutritional recommendations. We have introduced the concepts of ontology enrichment and population and the main motivations for its use in this project. We have presented the process by which our nutritional ontology is enriched with the concepts of the XML Schema diet model through an XSLT transformation to the OWL diet model. In this process the XSLT transformation and mapping rules were designed from scratch because none of the available tools fulfilled our requirements. The ontology is populated on-demand with the contents of a single diet, being unpopulated when it is no longer needed, storing the personalized diets as new XML documents in the repository. The inclusion of the diets and their components into the ontology through the enrichment and population processes elevates the syntactic level of the XML diets representation to the semantic level of the OWL ontologies allowing the reasoning over the contents of the diet in the knowledge base for performing the nutritional recommendations. The configuration of the NutElCare architecture, connecting this XML diet repository through a seamless and lightweight linkage to the ontologies of the knowledge base keeps the system timely responsive.

As future work, and because the number of diets in the system is growing significantly, we plan to improve the indexation of diets by taking advantage of the semantic classification in the ontology, in order to build more efficient indexes. Also, further work is required for the explicit mapping with accepted vocabularies containing concepts of the nutritional domain in the Linked Open Data.

Acknowledgements. This work was partially funded by the Innovation Office from the Andalusian Government under project TIN-6600 Virtra-el and by the 'Programa de Fortalecimiento de I+D+i' de la Universidad de Granada 2014–2015.

References

1. Baader, F., Horrocks, I., Sattler, U.: Description logics as ontology languages for the semantic web. In: Hutter, D., Stephan, W. (eds.) Mechanizing Mathematical Reasoning. LNCS (LNAI), vol. 2605, pp. 228–248. Springer, Heidelberg (2005)
2. Bermudez-Edo, M., Hurtado, M.V., Noguera, M., Hurtado-Torres, N.: Managing technological knowledge of patents: HCOntology, a semantic approach. Comput. Ind. (2015)
3. Bohring, H., Auer, S.: Mapping XML to OWL ontologies. In: Leipziger Informatik-Tage, vol. 72, pp.147–156 (2005)
4. Bühmann, L., Lehmann, J.: Universal OWL axiom enrichment for large knowledge bases. In: ten Teije, A., Völker, J., Handschuh, S., Stuckenschmidt, H., d'Acquin, M., Nikolov, A., Aussenac-Gilles, N., Hernandez, N. (eds.) EKAW 2012. LNCS, vol. 7603, pp. 57–71. Springer, Heidelberg (2012)
5. Buitelaar, P., Cimiano, P. (eds.): Ontology learning and population: bridging the gap between text and knowledge, vol. 167. Ios Press, Amsterdam (2008)
6. Caracciolo, C., Stellato, A., Morshed, A., Johannsen, G., Rajbhandari, S., Jaques, Y., Keizer, J.: The Agrovoc linked dataset. Semant. Web **4**(3), 341–348 (2013)
7. Espín, V., Hurtado, M.V., Noguera, M.: Towards holistic support of active aging through cognitive stimulation, exercise and assisted nutrition. In: Pecchia, L., Chen, L.L., Nugent, C., Bravo, J. (eds.) IWAAL 2014. LNCS, vol. 8868, pp. 312–319. Springer, Heidelberg (2014)
8. Espín, V., Hurtado, M.V., Noguera, M., Benghazi, K.: Semantic-based recommendation of nutrition diets for the elderly from agroalimentary thesauri. In: Larsen, H.L., Martin-Bautista, M.J., Vila, M.A., Andreasen, T., Christiansen, H. (eds.) FQAS 2013. LNCS, vol. 8132, pp. 471–482. Springer, Heidelberg (2013)
9. Ferdinand, M., Zirpins, C., Trastour, D.: Lifting XML schema to OWL. In: Koch, N., Fraternali, P., Wirsing, M. (eds.) ICWE 2004. LNCS, vol. 3140, pp. 354–358. Springer, Heidelberg (2004)
10. Fernández, M., Cantador, I., López, V., Vallet, D., Castells, P., Motta, E.: Semantically enhanced information retrieval: an ontology-based approach. Web Semant. Sci. Serv. Agents World Wide Web **9**(4), 434–452 (2011)
11. Ghawi, R., Cullot, N.: Building ontologies from XML data sources. In: DEXA Workshops, pp. 480–484 (2009)
12. Hacherouf, M., Bahloul, S.N., Cruz, C.: Transforming XML documents to OWL ontologies: A survey. J. Inf. Sci. **41**(2), 242–259 (2015)
13. Heckmann, D., Schwartz, T., Brandherm, B., Schmitz, M., von Wilamowitz-Moellendorff, M.: Gumo–the general user model ontology. In: Ardissono, L., Brna, P., Mitrovic, A. (eds.) User modeling 2005. LNCS, vol. 3538, pp. 428–432. Springer, Heidelberg (2005)
14. Hitzler, P., Krotzsch, M., Rudolph, S.: Foundations of Semantic Web Technologies. CRC Press (2011)
15. Hitzler, P., Krötzsch, M., Parsia, B., Patel-Schneider, P.F., Rudolph, S.: OWL 2 web ontology language primer. W3C Recommendation, 27 October 2009. http://www.w3.org/TR/owl2-primer/
16. Horridge, M., Bechhofer, S.: The OWL API: a java API for OWL ontologies. Semant. Web **2**(1), 11–21 (2011)

17. Petasis, G., Karkaletsis, V., Paliouras, G., Krithara, A., Zavitsanos, E.: Ontology population and enrichment: state of the art. In: Paliouras, G., Spyropoulos, C.D., Tsatsaronis, G. (eds.) Knowledge-Driven Multimedia Information Extraction and Ontology Evolution. LNCS, vol. 6050, pp. 134–166. Springer, Heidelberg (2004)
18. Sirin, E., Parsia, B., Grau, B.C., Kalyanpur, A., Katz, Y.: Pellet: a practical owl-dl reasoner. Web Semant. Sci. Serv. Agents World Wide Web 5(2), 51–53 (2007)
19. XSD2OWL. http://rhizomik.net/html/redefer/#XSD2OWL

Automatic Generation of CVs from Online Social Networks

Sergio Maia Dias[1], Alda Lopes Gancarski[2]([✉]), and Pedro Rangel Henriques[1]

[1] Department of Computer Science, CCTC, University of Minho,
Campus de Gualtar, Braga, Portugal
pg25338@alunos.uminho.pt, prh@di.uminho.pt
[2] Institut Telecom, Telecom SudParis, CNRS SAMOVAR,
9 rue Charles Fourier, 91011 Évry, France
Alda.Gancarski@telecom-sudparis.eu

Abstract. Since the explosion of Social Media use, users information being dessiminated and dynamically updated, Curriculum Vitae (CV) documents started to be automatically generated, compiling that information and returning it to the user usually in PDF file format. However, existing CV generation tools do not use a standard CV structure format, which should be generic enough for common user needs, but also with domain specific components for certain work environnements, like academic and research. Another difficulty on using most of existing tools is that they return CV in a printable file format, not easily editable. In this paper, we introduce *CVGenie*, a system to automatically generate CV from information available in Online Social Networks. The system uses the EuroPass CV standard, extended with domain specific components. The CV file format is the XML dialect of EuroPass, because not only it is editable, but also it allows for the interoperability with other applications.

1 Introduction

In a professional environment, individuals need a way to expose their employment history, qualifications and education, to prove their worth in a competitive marketplace. In an academic environment, individuals also have this requirement, although it is tailored to the nature of the individuals' work, i.e., more focused in their research and teaching.

Previously, this requirement was satisfied by a *curriculum vitae* (CV), which is a document that contains an overview of the aforementioned aspects of an individual's career. With the emergence of the web, those aspects started to be exposed in personal or institutional web pages.

As such, with the popularity of social media, professional-oriented Online Social Networks (OSN) started to emerge, and became essential for personal promotion in the labor market. In these social networks, users share their professional experience and the projects (thematics and collaborators) they were or are involved with. An example of these OSN is *LinkedIn* [4], where users enumerate the companies or institutions that they have worked in the past, they list

© Springer International Publishing Switzerland 2015
J.-L. Sierra-Rodríguez et al. (Eds.): SLATE 2015, CCIS 563, pp. 258–263, 2015.
DOI: 10.1007/978-3-319-27653-3_25

the projects that they have been involved in and they describe their education and create an explicit network of past or present collaborators. Other examples, in a more academic focused social networks, are *Academia* [1] and *ResearchGate* [5], where the users' publications and projects are displayed.

Nonetheless, this method of self-marketing is insufficient for some processes; to apply for a position at a company it is usually required for the candidate to present his professional experience in a succinct manner, ordinarily in a document for that effect, i.e., a CV. It is thus clear that there is a need to utilize the information available in these new forms of professional exposure for the automatic generation of CV.

This paper describes an ongoing work which entails the creation of *CVGenie*, a tool that, using the information available online on specific social networking platforms, like *LinkedIn* [4] or *Academia* [1], generates CVs for the end user, who can then use them at need.

This paper is organised as follows: Sect. 2 introduces existing CV formats. Existing systems that perform CV generation are analysed in Sect. 3. Section 4 describes the proposed system, showing its main features and architecture. We finish the paper with a conclusion and the planned future work.

2   Curriculum Vitae Standard Formats

Nowadays, several CV formats exist, defined by different institutions or countries. Despites there is not a standard globally accepted, the *EuroPass* format [3] comes close: it is an initiative by the Directorate-General for Education and Culture of the European Union to standardize CV documents, and is widely accepted in several countries of the European Union, like Portugal and Spain. Other CV formats are more focused on specific areas of expertise, such as the academic format of The Career Center of the University of Washington (UW's Academic CV) [2], or the College Art Association's Visual Artist Format (CAA's Artist CV) [6]. Since these formats are tailor made for specific individuals or a specific situation, they are not universally accepted. Moreover, they are also not very strict, since they do not define a clear set of rules that a CV must comply with, delegating that responsibility to the individual writing the document.

CV formats vary according to the individuals' context (geographical location, area of expertise) and the needed detail level. However, a core set of information is common among most formats, as shown in Table 1.

UW's Academic CV and CAA's Artist CV are examples of formats dedicated to a specific work domain. In its turn, *EuroPass* is a popular generic-purpose format used by individuals in a wide range of areas of expertise, this is why it is our system's CV format. Domain specific content and the desired level of detail will be integrated in *EuroPass* through adequate extensions.

Table 1. Common information for existing formats

	EuroPass	UW's Academic CV	CAA's Artist CV
Name	Yes	Yes	Yes
Gender	Yes, optional	No	No
Date of birth	Yes, optional	No	Yes, optional
Addresses	Yes, optional	Yes	Yes, optional
Contacts	Yes	Yes	Yes
Education	Yes	Yes	Yes
Work experience	Yes	Yes	Yes
Skill set	Yes	Yes	Yes
Affiliations	Yes, optional	Yes, optional	Yes, optional
Seminars	Yes, optional	Yes, optional	Yes, optional
Publications	Yes, optional	Yes	Yes
References	Yes, optional	Yes, optional	Yes, optional

3   Existing Solutions for CV Generation from Online Social Networks

Several applications (software packages) exist for CV generation from OSN, like Yevgeniy Brikman's *Resume Builder* [12], *DoYouBuzz.com* [9], *VisualCV* [13], *Create-CV.com* [7] and EGrabber's *ResumeGrabber Suite* [11].

These applications are capable of extracting information from OSN, being *LinkedIn* the most common source of data. Some, like *Create-CV.com*, even support other social networking and messaging services, like *Facebook*, *Twitter*, *Skype*, etc., although most of them are either poor sources of professional information or require extensive mechanisms for extracting data, which makes them not feasible.

The main objective of some of these applications is to collect the information and make it available online with a structure commonly associated with CV, with the option of further editing it with the integrated tools, and possibly export it as a file ready for printing. This is the case of *Resume Builder*, *DoYouBuzz.com* and *VisualCV*. *Create-CV.com* simply allows to export the end result as a file, and doesn't allow displaying it online. These solutions focus on being complete and self-sufficient, once the information as been retrieved from the data sources. In doing so they become less useful, because they do not interoperate with other systems, not being able to integrate different complementar functionalities. In addition, generated files by those solutions cannot be reused by other systems easily. For example, none of these tools generates files in a file format that is easily editable externally, or that follows a standard like *Europass* (which uses an XML based format) and therefore can be integrated with other systems.

EGrabber's *ResumeGrabber Suite* is a substantially different solution, since it is not focused on the individual described in the CV as the end user, but is a tool

Table 2. Comparison of features of the identified CV generation systems

	Resume builder	DoYouBuzz	VisualCV	Create CV
Import info from online media	Yes, from LinkedIn	Yes, from LinkedIn and Viadeo	Yes, from LinkedIn	Yes, from LinkedIn, Facebook, Twitter, Skype, etc. (although it only retrieves basic information)
Import info from files	No	No	Yes, from PDF and Word	No
Export CV in read-only file format	Yes, in PDF	Yes, in PDF, although it only exports the basic information	Yes, in PDF	Yes, in PDF
Export CV in editable file format	No	Yes, in Word, although it only exports basic information	No	No
Has an integrated editor	Yes	Yes, with many features	Yes, with many features	No
Can share CV online	Yes, through a link	Yes, through a link, social networks or publicly on search engines	Yes, through a link, social networks or publicly on search engines	Yes, through a link
Includes info for specific areas of expertise	No	No	No	No

to be used in the context of an organization for integration of CV information of the organization's collaborators or job applicants.

Table 2 depicts the features of each tool, establishing a comparison and identifying the main issues that were detected. From Table 2, we see that the main objective of these systems is to collect the individual's data, allow for it to be easily edited inside the system and easily share it online, or export it as a read-only format. The biggest flaws in these systems are:

- In general, they do not allow exporting the CV in an editable format.
- In general, they only import information from the most popular source of professional data, i.e., *LinkedIn*.
- They just include in the CV the core set of information identified in Table 1; more specific one, like detailed academic information, is not considered.

These aspects are considered in the proposed CVGenie system.

4   CVGenie System: Requirements and Architecture

The system projected will be designed towards users that want to generate their own CVs, from the information about themselves available online. The features that shall be provided by that system are the following:

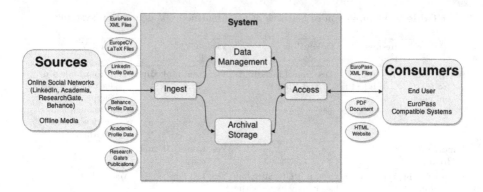

Fig. 1. The proposed system architecture

- **Import Information from OSN:** The information required for the CV will be obtained from the selected online social networks (*LinkedIn*, *Academia*, *ResearchGate* and *Behance*). The latter will be used mainly for information for the academic and artistic sections of the CV.
- **Import Information from Files:** The system will be able to import information from files in the standard EuroPass XML format as well as LaTeXfiles that use the *EuropeCV* [8] package.
- **Export CV in an Editable Format:** The system will be able to export CV documents in the EuroPass XML format, making it possible to later import these documents in EuroPass compatible systems.
- **Export CV in a Read-only Format:** The system will be able to export CV documents as PDF files or simple HTML websites.
- **Include Information for Specific Areas of Expertise:** Extensions to EuroPass will be proposed to support specific formats like UW's Academic CV and CAA's Artist CV.

The proposed CVGenie architecture, presented in Fig. 1, is similar to the OAIS model [10]; it will provide the same standard features, like information ingestion, data management and storage, and knowledge dissemination.

The sources of data for the our CVGenie tool will be offline media, which includes both EuroPass XML documents and LaTeX documents using the EuropeCV package, and online media from user profiles in various online social networks such as LinkedIn, Behance, Academia and Research Gate.

This information will be ingested by the system: it will be interpreted and stored.

Afterwards, the user can then export the information in an editable format, which will be the EuroPass XML format, or in a read-only format, which will include both PDF documents and simple HTML websites.

5   Conclusion and Future Work

The system we propose in this paper is dedicated to automatic CV generation using user's information from OSN. Our system is intended to extract information from the most popular OSN, while returning the CV information in a generic standard format that can support domain specific extensions if needed. The standard CV structure adopted is EuroPass, in its XML format, but other file formats can be produced, like PDF or HTML.

Currently it is possible to import and export CVs in the EuroPass XML format. It is also possible to import CV data from LinkedIn, taking into account both the existing data on the user's CV as well as the available data on LinkedIn.

When importing data from LinkedIn, a comparison is performed, and if conflicting data is found, the user's intervention is requested: a page is displayed, listing the fields in conflict, and the user can choose, field by field, which data should be kept on his CV. This mechanism will be featured when importing data from other OSNs in future implementations. Data available to be imported from LinkedIn is most basic information, including: first and last name, location and last active job position. More detailed information is not available due to limited access to the LinkedIn API.

As future work, we intend to perform the following tasks:

- Formally define extensions to the EuroPass format dedicated to specific domains.
- Make large and rigorous system validation and evaluation with users from different interest domains having accounts on different OSN.

Acknowledgment. This work is co-funded by the North Portugal Regional Operational Programme, under the National Strategic Reference Framework (NSFR), through the European Regional Development Fund (ERDF), within project GreenSSCM - NORTE-07-02-FEDER-038973.

References

1. Academia.edu - Share research (2014). www.academia.edu
2. Academic Careers Curriculum Vitae - The Career Center of the University of Washington (2014). http://careers.washington.edu/ifiles/all/files/docs/grad students/pdfs/AcademicCareers-Curriculum_Vitae_07-08.pdf
3. Europass: Curriculum Vitae (2014). http://europass.cedefop.europa.eu/en/ documents/curriculum-vitae
4. LinkedIn: World's Largest Professional Network (2014). www.linkedin.com
5. ResearchGate (2014). www.researchgate.net
6. Standards and Guidelines — College Art Association — CAA (2014). http://www. collegeart.org/guidelines/visartcv
7. Create my CV online for free (2015). Create-CV.com
8. CTAN: Package europecv (2015). http://www.ctan.org/pkg/europecv
9. DoYouBuzz: Your best resume (2015). DoYouBuzz.com
10. ISO 14721:2012 - Space data and information transfer systems – Open archival information system (OAIS) – Reference model (2015). http://www.iso.org/iso/ home/store/catalogue_ics/catalogue_detail_ics.htm?csnumber=57284
11. Software to Import Resumes (2015). www.egrabber.com/resumegrabbersuite
12. Turn your LinkedIn Profile into a Resume — Resume Builder (2015). resume.linkedinlabs.com
13. VisualCV - Online CV Builder and Professional Resume CV Maker (2015). VisualCV.com

Knowledge Identification from Requirements Specification

Eduardo Barra[⊠] and Jorge Morato

Universidad Carlos III de Madrid,
Avda. Universidad 30, 28911 Leganés, Madrid, Spain
ebarra@kr.inf.uc3m.es, jmorato@inf.uc3m.es

Abstract. One of the main artifacts in Requirements Engineering is the Requirements Specification (RS). Throughout the life cycle of the RS arises the need of extracting knowledge in order to facilitate communication with stakeholders. However, this process is not usually efficient. In the different proposals for the representation of an RS conflicts often arise, due to coupling and redundancy of requirements. The Aspect-Oriented paradigm provides principles to address a multidimensional modeling to avoid these conflicts. Knowledge-Engineering is proposed to provide a model of the knowledge needed to allow an efficient extraction of knowledge from the requirements by ontologies. An experimental study has been developed to assess its efficiency when compared with classical methods.

Keywords: Requirements Engineering · Requirements Specification · Aspect-Oriented · Knowledge-Engineering · Ontology · Knowledge extraction

1 Introduction

The most important artefact to be used to transmit knowledge effectively between the different activities of any process of Requirements Engineering (RE) is the Technical Specification Document (TSD). This document is written in natural language, and is often supplemented with graphical models for its better understanding. The Requirements Specification (RS) is the section of a TSD that contains the knowledge specification of a software product.

The internal organization of an RS is aimed at reducing the complexity of its semantics and improving the global understanding of the requirements for the effective transmission of knowledge. However, there is wide agreement on organizing requirements in a simplistic way. In fact, they are frequently classified in just two types: functional and non-functional. In this convention the group of functional requirements is accepted by most analysts, however, the organizing of non-functional requirements is ambiguous and subjective. Different researchers claim that, in fact, they impose restrictions on the functional requirements [1]. Therefore, they should be viewed as properties of future software products [2] or as requirements that indicate quality [3]. We think that the paradigm Aspect-Oriented (AO) [4] provides the resources to develop these models that allow a multidimensional organization without compromising the integrity of its semantics. Therefore, the representation of an RS with this paradigm

© Springer International Publishing Switzerland 2015
J.-L. Sierra-Rodríguez et al. (Eds.): SLATE 2015, CCIS 563, pp. 264–270, 2015.
DOI: 10.1007/978-3-319-27653-3_26

reduces the conflicts, couplings and redundancy between requirements. This reduction in the complexity of a software product facilitates the extraction of knowledge, its reuse and management.

The discipline Ontology Engineering is often applied to Knowledge Engineering (KE). Ontologies allow to create an explicit and formal specification of knowledge-managed through computers that provide reusability and shareability [5, 6]. Therefore, our hypothesis is that the representation of an RS through ontologies is an important factor to obtain an efficient representation of knowledge. With this background, this research proposes guidelines according to the AO paradigm based on principles of KE applied to Requirements Engineering (RE) for the development of an RS. OE provides a multidimensional semantic model with a well-founded organization that facilitates the extraction of knowledge.

In accordance with these ideas, the rest of the paper is structured as follows. Section 2 briefly discusses related work for the modeling of an RS based on the paradigm AO and OE. In Sect. 3, some guidelines for the modeling of an RS are proposed and exemplified. In Sect. 4, an evaluation is developed to demonstrate that the representation with the guidelines is effective and efficient in the extracting of knowledge from RS. Finally, Sect. 5 presents the conclusions and future work.

2 Related Work

Our study is focused on the early stages of the software development process (SDP). SDP is usually split into different stages. In the AO paradigm, we can talk about early, middle and late aspects, where the early aspects comprise the requirement specification and the architectonic logical representation. Late aspects deal with the low codification level, and finally, middle aspects are located between these stages (Table 1).

Table 1. Stages in different software development proposals

Activity	USDP	MDA	AO
Specification	Requirements	CIM	Early aspects
	Analysis		
Development	Design	PIM	
		PSM	Middle aspects
	Implementation		Late aspects

In recent years, many studies have shown the potential of the AO paradigm focused on the early stages of software development. These proposals are grouped under the term Aspect-Oriented Requirement Engineering (AORE) [7, 8]. The first step in this research has been to study the most representative approaches in AORE, in order to find out well-founded approaches to include in our work.

In 2003, Rashid et al. proposed using XML language to index the description of the requirements for the management of concerns [9]. In another proposal, "Multi-Dimensional Separation of Concerns in Requirements Engineering" [10], the suggested solution for RE is the separation of concerns into multidimensional categories, in order to provide a categorization modeled from different points of view. In Yu and Prado's work is [11] proposed aspect-orientation techniques to manage objectives. It suggests mechanisms to avoid conflicts between requirements. Another proposal, that combines aspect-oriented analysis and design [12], highlights the need to decompose the requirements into more elementary parts. Finally, the research carried out by Jacobson and Ng shows that when the concepts of the AO paradigm are related to "use cases" the positive influence for identifying concerns can be observed [13].

In these proposals, not only did we find different contributions to be considered, but also different problems. These approaches only consider the modelling of requirements in descriptions of natural language, where a requirement may belong to different aspects.

The modelling of requirements in these proposals lacks a clear technique for the modelling of knowledge, generating conflict, coupling and semantic redundancy. AORE proposals are aimed at the identification, separation and composition of concerns but they are not oriented to the representation of knowledge in an efficient way.

Additionally, different ontology-driven approaches to support RE have been analyzed [14–17]. Although modeling the knowledge of the domain through ontologies addresses an obvious need, there is a lack of well-established natural language techniques to represent these internal semantic relationships among requirements. Besides, the need of developing organizational structures to represent the Requirements Specification (RS) is usually overlooked.

3   Guidelines for the Semantic Modeling of an RS

The first stage of the guidelines proposed in this work focused on the importance of a sound and adequately substantiated organization in the development of an RS. In this case, we propose the creation of a structure to specify the requirement. This structure is designed in accordance with different viewpoints. The main goal is to provide a way for the development of an RS that improves the understanding of the requirements as a whole. In this regard, we have proposed the use of an Architecture Viewpoint (AVP) that provides a schema for developing the RS. The AVP is modeled on a domain ontology that provides a guide for modelling responsibilities. The aim is to identify the viewpoints that will group both the dominant concerns and the related viewpoints in crosscutting concerns. The high-level viewpoints proposed in the AVP will act as containers of viewpoints of lower level in a recursive nesting to reach the last level viewpoint.

In the second stage, the early concerns of a software product are modeled according to the AVP base, in an organized knowledge representation. The goal of these guidelines is to model the early knowledge using ontologies, and leaving as a secondary activity elaborating a description in natural language, in order to facilitate the understanding by non-expert stakeholders. The modelling of concerns through the Ontology Engineering has its major support in the Ontology Web Language (OWL). This is a recommendation introduced in 2004 by the W3C for building the Semantic Web, which is the most popular language for the semantic description of ontologies. Accordingly, OWL has been the language selected in these guidelines for the modelling of an RS. The application of the guidelines involves modeling of knowledge of the Requirements, typical of an RS with the specification of the properties between concerns. There are two main types of properties that OWL represents as relationships: "Object Properties" and "Data Properties". The most important relationships in these guidelines are the "Object Properties" to model the semantic relationships. The modelling of early knowledge of a software product assisted by these guidelines involves the use of natural language in order to allow the developer a richer description of the concerns, in the same way that it is done with typical requirements. This information is added to the "Annotation Properties".

An efficient representation of a specification requires split its components in elements like concerns, information entities, roles, conditions, and so on. Usually specifications are expressed in natural language (NL). Unfortunately NL is complex and prone to ambiguities. An example may explain some of these difficulties of modeling processes in an RS. In Fig. 1 the decomposition of concern expressed in a NL is shown. A multidimensional modeling to split its components in a well-grounded way is necessary to avoid non-atomic and overlapped requirements.

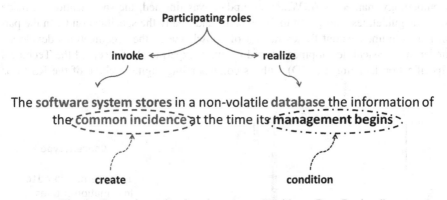

Fig. 1. Constraints related to the concern "Incidence Data Register"

4   Evaluation

In order to show the efficiency of the guidelines for the RS creation, we developed an experiment to compare our guidelines with a classical development. The case study proposed was to develop an application to manage a "stadium for athletics events".

At the step 1 of the experiment, a solution was developed with the method described (hereafter referred to as specification B). In addition, we gave to 32 teams a description of the case study. The project was given to students in Software Engineering in the final year of the degree in computer science. Each team was comprised of 5 members. The teams had to specify the TSD of a software product under a classical development process. Teams were allowed to base their solution on IEEE 830 or ESA PSS-05 standards.

The specifications developed under a classical methodology were analyzed by five researcher teachers, all of them experts in RE. The experts selected the two best ones, which we call specifications A and C for short.

The second step of the experiment involved to extract shared crosscutting-concerns, dominant-concerns and common information entities from the different solutions. Due to its importance in this proposal, the classification of crosscutting concerns is focused on the type of query. The category "information entity" comprises the name of classes and view points, while information related to information entities refers to class properties and its values. Thus, the concerns were identified for the experiment referee on every of the three solutions developed.

At third step, we asked to six software analysis experts from different software factories to manage the three specifications when facing a major update in the software. Therefore they were required to obtain the same knowledge from the three specifications. In order to avoid a biased result due to the learning of the domain, the specifications were given to the expert in a specific order. In so doing, the first specification given is not at a disadvantage in terms of effectiveness of change when comparing with the other specifications. The first specification given was developed with the classical methodology, named as A. When the update was finished, the specification B, made with the guidelines, was given to the experts, and finally the specification C. In the part of the experiment about the extraction of knowledge of the specifications developed under the classical development A and C, every expert analyst received the Technical Specification Document (TSD) with its corresponding digital archive of the RS to be

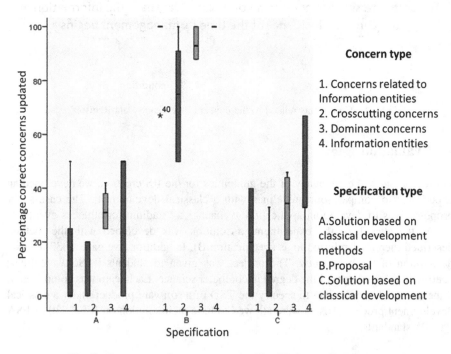

Fig. 2. Percentage of correct concerns identified according its type

used in a management of requirements. In the part of the experiment about the extraction of knowledge of the specifications developed under our guidelines, the six expert analysts received an OWL archive. Tools for ontologies, such as editors, reasoners and a search engine were provided to the experts to carry out the update. The time to work in every specification was limited to 30 min.

At the step 4, results have been compared and analyzed. The evaluation was made quantifying the concerns to update correctly identified by the experts analysts from the total knowledge proposed in each specification.

The result is shown in Fig. 2. We can clearly see that the specification of a software product on the early stages developed with the guidelines allows the knowledge extraction more efficiently.

5 Conclusions and Future Works

The first proposal of the presented approach is to use the AO paradigm concepts to get a multidimensional modeling. This step allows us to avoid conflicts, such as couplings and redundancy in representing RS. These conflicts appear in a recurrent way in modelling RS under classical methodologies. An additional goal is to prevent problems due to ambiguities when representing in natural language RS. In this sense, the KE concepts allow to solve the problem about the knowledge representation in the early stages of a software product.

Several approaches exist for RS modeling on AO and KE. They typically lack knowledge organization and structure. Therefore, they are neither efficient nor effective in defining the knowledge that different stakeholders require.

One of the contributions of our approach is the proposal of guidelines that improve the RE process, creating an RS model that facilitates the knowledge extraction. The guidelines are based on the improvement of the semantic representation of the requirements, with a multidimensional knowledge organization.

In this methodology, regular updates are properly incorporated thanks to the way we split the requirements and user stories. For this reason, the proposal is well-suited to work with agile methods for managing product development.

We developed an experiment with the proposed guidelines to study the effectiveness of the approach. The result of this evaluation proved that the innovative solution proposed by this work improves substantially the knowledge extraction in the early stages avoiding investing many resources trying it.

As a future work we are developing guidelines for the creation of the AVP in a systematic way, which allows us to structure and organize those aspects that support the evolution of a software product. In this regard, the automation of the guidelines will provide the support for the creation of the AVP needed by the software factories, where the resultant AVP could be reused in the same line product. To complement this work, we intend to create a tool to manage concerns. This tool must provide the resources that a typical tool offers to manage requirements, but respecting the concepts of the conceptual AO model in a clear way. That is, support will be given through a virtual guide to describe the requirements, but also modelling the attributes with the necessary concerns that describe their knowledge.

References

1. Sommerville, I.: Software Engineering. Pearson Education, Boston (2005)
2. Jacobson, I., Booch, G., Rumbaugh, J.: The Unified Software Development Process. Addison-Wesley, Reading (1999)
3. Doerr, J., Kerkow, D., Koenig, T., Olsson, T., Suzuki, T.: Non-functional requirements in industry – three case studies adopting and experience-based NFR method. In: Proceedings of the 13th IEEE International Conference on Requirements Engineering, pp. 373–382. IEEE, New York (2005)
4. Kiczales, G., Lamping, J., Mendhekar, A., Maeda, C., Lopes, C.-V., Irwin, J.: Aspect-oriented programming. In: Akşit, M., Matsuoka, S. (eds.) ECOOP 1997. LNCS, vol. 1241. Springer, Heidelberg (1997)
5. Berners-Lee, T., Hendller, J., Lassila, O.: The semantic web. Sci. Am. **284**(5), 29–37 (2001)
6. Brewster, C., O'Hara, K.: Knowledge representation with ontologies: the present and future. IEEE Intell. Syst. **19**(1), 72–81 (2004)
7. Grundy, J.: Aspect-oriented requirements engineering for component-based software systems. In: IEEE International Conference on Requirements Engineering, p. 84. IEEE, New York (1999)
8. Rashid, A., Sawyer, P., Moreira, A.M.D., Araújo, J.: Early aspects: a model for aspect-oriented requirements engineering. In: Proceedings International Conference on Requirements Engineering, pp. 199–202. IEEE, New York (2002)
9. Rashid, A., Moreira, A., Araújo, J.: Modularisation and composition of aspectual requirements. In: Proceedings of the 2nd International Conference on Aspect-Oriented Software Development, pp. 11–20. ACM, Boston (2003)
10. Moreira, A., Rashid, A., Araujo, J.: Multi-dimensional separation of concerns in requirements engineering. In: 13th IEEE International Conference on Requirements Engineering, pp. 285–296. IEEE, New York (2005)
11. Yu, Y., do Prado Leite, J.C.S., Mylopoulos, J.: From goals to aspects: discovering aspects from requirements goal models. In: 12th IEEE International Requirements Engineering Conference, pp. 38–47. IEEE, New York (2004)
12. Baniassad, E., Clarke, S.: Theme: an approach for aspect-oriented analysis and design. In: Proceedings of the ICSE 2004, pp. 158–167. IEEE, Washington (2004)
13. Jacobson, I., Ng, P.-W.: Aspect-Oriented Software Development with Use Cases. Addison-Wesley, New Jersey (2005)
14. Kaiya, H., Saeki, M.: Using domain ontology as domain knowledge for requirements elicitation. In: 14th IEEE International Requirements Engineering, pp. 189–198. IEEE, Los Alamitos (2006)
15. Jureta, I.J., Mylopoulos, J., Faulkner, S.: Revisiting the core ontology and problem in requirements engineering. In: 16th IEEE International Requirements Engineering, RE 2008, pp. 71–80. IEEE, Los Alamitos (2008)
16. Velasco, J.L., Valencia-Garcia, R., Fernandez-Breis, J.T., Toval, A.: Modelling reusable security requirements based on an ontology framework. J. Res. Pract. Inf. Technol. **41**(2), 119–133 (2009)
17. Souag, A., Salinesi, C., Wattiau, I., Mouratidis, H.: Using security and domain ontologies for security requirements analysis. In: Computer Software and Applications Conference Workshops (COMPSACW), 2013 IEEE 37th Annual, pp. 101–107. IEEE, Los Alamitos (2013)

Author Index

Printed in the United States
By Bookmasters